The American Synthetic Rubber
Research Program

D0087445

The Chemical Sciences in Society Series

Arnold Thackray, Editor

The American Synthetic Rubber Research Program

Peter J. T. Morris

upp

UNIVERSITY OF PENNSYLVANIA PRESS Philadelphia

Copyright © 1989 by the University of Pennsylvania Press
Printed in the United States of America

Permission is granted by R. D. Ulrich to reprint two graphs from
*Contemporary Topics in Polymer Science, Volume 1: Macromolecular
Science—Retrospect and Prospect*, edited by R. D. Ulrich, New York: Plenum
Press, 1978, p. 3.

Jacket illustration reprinted by permission of Charles Scribner's Sons, an
imprint of Macmillan Publishing Company, from *Man in a Chemical World*
by A. Cressy Morrison. Copyright 1937 Charles Scribner's Sons; copyright
renewed © 1965 Betty Morrison Bond.

Library of Congress Cataloging-in-Publication Data

Morris, Peter John Turnbull.
    The American synthetic rubber research program / Peter J. T.
Morris.
        p.   cm. — (The chemical sciences in society series)
    Bibliography: p.
    Includes index.
    ISBN 0-8122-8205-1
    1. Rubber, Artificial—Research—United States.   I. Title.
II. Series.
    TS1925.M58   1989
678'.72'072073—dc20   89-40399
                                                              CIP

This book is dedicated to the memory of

GLADYS JONES VALE (1907–1988)

"I was a stranger, and ye took me in"
R.I.P.

Contents

Foreword by Arnold Thackray   ix
Acknowledgments   xi
Introduction   1

Chapter One
The U.S. Government Rubber Program

On the Banks of the Rhine: 1929–1936   7
From the Rhine to the Cuyahoga: 1936–1941   8
The Emergency Program: 1940–1942   9
Controversy on the Potomac   10
Research Program Established   12
Creation of a New Industry: 1942–1945   13
The Postwar Period: 1945–1950   16
The Korean War: 1950–1953   17
The Disposal of the Plants: 1953–1955   19
The End of the Research Program: 1955–1957   20
Since the Program: 1955–1985   22

Chapter Two
Innovation in the Synthetic Rubber Industry

Introduction   27
The Corporate Context   28
The Wartime Research   29
Cold Rubber and Oil-Extended Rubber   32
Intermezzo: Infra-Red Spectrophotometry   39
Synthetic "Natural Rubber"   41

Butyl Rubber   49
Success or Failure?   50

Chapter Three
The Universities

Introduction   69
Prior Industry-University Collaboration   69
Illinois: Chemical Structure and Organic Synthesis   70
Chicago: Free Radicals and Emulsions   75
Minnesota: Analysis and Kinetics   79
Cornell: Physical Methods and Polymer Configurations   83
The Polymer Research Discussion Group   86
The Universities and Industrial Innovation   88

Chapter Four
Quality Control and Polymer Evaluation

Introduction   101
Bell Telephone Laboratories   102
National Bureau of Standards   109
Government Laboratories   112
Generic Research in the Rubber Program   115

Chapter Five
The Rubber Research Program and Polymer Science

Introduction   125
The Spin-Offs   127
Flory and the Rubber Research Program   132
Polymer Education   133
The Polymer Research Discussion Group Revisited   134
A Missed Opportunity?   136

Summary of Major Conclusions   141

Appendix: Introduction to Polymer Chemistry   143
   Select Bibliography   161
Works Cited   163
Subject Index   179
Name Index   182
Polymers and Monomers Index   190

Foreword

The Chemical Sciences in Society: what may one expect in a series with such a title? The question is legitimate. The answer is unsurprising, but not unimportant.

Science and technology have transformed our world and our ideas. Our physical environment, our material culture, our conceptual systems, and our manner of living have been changed, are being changed, and will be changed by science. That much is familiar. Less noticed is the way in which, of all the scientific and technological domains, those associated with the chemical sciences have most often been ignored or misunderstood—subject to ignorance or fear. The very word "chemical" has become synonymous with "undesirable"—as in "chemical dependency." Chemistry is seen as somehow the cause of acid rain, air pollution, carcinogens, drugs, environmental problems, ozone depletion, and smog, to offer only an abbreviated list. At the same time, the chemical sciences are often viewed as intellectually unimportant—the mere detail that fills in the lofty work of physics, the fine print that obscures the grand designs of biology.

The aim of *The Chemical Sciences in Society* is to redress this balance, by bringing into prominence the growing body of scholarship which testifies to the importance of the chemical sciences in society. Chemistry has always been the most earthy and the most central of the sciences. Its earthiness—its connection to the colors, smells, and sounds of substance and of change—stretches backward into the mists of alchemy. But its earthiness also extends into the huge, complex business enterprises of our day, in petrochemicals, in pharmaceuticals, in the cornucopia of polymers and plastics, in composites and "advanced materials," in agriculture and biotechnology, and in cryogenics and electronics. Likewise, the centrality of the chemical sciences may be seen in their role in areas as varied as molecular biology, materials science, and the clinical sciences.

Also worthy of note is the way in which science and technology are increasingly to be understood as political. Science today is an arena in which actors compete for scarce resources—of prestige, of intellectual turf, of government support, of student interest, of corporate involvement, and of public acclaim. The chemical sciences offer a particularly rich area of study when seen in this way—thanks to their long history, their ubiquity, their commercial significance, and their involvement with government and medicine (no monarch could ignore someone able to transmute base metals into gold, or to offer the elixir of life—either in the Renaissance or in the National Institutes of Health).

The chemical sciences in society thus offer a rewarding field for serious study. The aim of the present series is to capture the best of the scholarship developing in this field, whether it deals with recent events or with more traditionally "historical" periods. Our desire is not to be partisan, but to be serious and balanced. The Beckman Center for the History of Chemistry is committed to helping bring into being a deeper, more reflective analysis of the chemical sciences. We trust *The Chemical Sciences in Society* will assist toward this end.

Arnold Thackray

Acknowledgments

The research for, and writing of, this book was greatly aided by financial support from the National Endowment for the Humanities. I thank the Endowment for its help, and Daniel P. Jones for his thoughtful interest. Additional funding came from the National Science Foundation, via the Polymer Project of the Arnold and Mabel Beckman Center for the History of Chemistry. In the closing stages of this book, I was funded by the Royal Society and the British Academy as a postdoctoral research fellow.

I am grateful to all my colleagues at the Beckman Center for their help: to Arnold Thackray for his invitation to join BCHOC for two and a half years, and for his able direction of this project; to Jeffrey Sturchio, who gave me moral support, advice, new leads to follow up, and a vast number of bibliographical references; to John Heitmann, who laid the groundwork for the project; and to James J. Bohning, Mary Ellen Bowden, Eric Elliott, Ray Fergusson, Colleen Wickey Mason, Herman Skolnik, and George Tselos.

Many veterans of the synthetic rubber research program assisted my research by giving interviews, advice, and new contacts. I would like to thank everyone who found time to talk about their experience of a program they all recall with pleasure and pride. A list of interviewees can be found at the beginning of "Works Cited." I am particularly grateful to Roger Beatty, James D'Ianni, Paul S. Greer, Maurice Morton, Charles C. Price, and Herman Schroeder for their advice and encouragement. The draft of this book was reviewed by William O. Baker, Attilio Bisio, James D'Ianni, Calvin S. Fuller, Paul S. Greer, Edward Meehan, and Charles C. Price. Their valuable advice is acknowledged in the notes. Frank McMillan read the section on the development of synthetic natural rubber, and sent me copies of his interview notes and documents.

My research was aided by several libraries and archives. John Miller at the University of Akron assisted my exploitation of the program research reports and the Goodrich archives. Ruth Murray at the ACS Rubber Division office in the Gifford Library at the University of Akron was a pillar of strength: contacting program participants, finding obscure articles, and more recently, speedily answering frantic requests for information faxed from England. I would also like to thank the staff of the National Archives (Economic and Social Branch and Modern Military Branch), AT&T Bell Telephone Laboratory Archives (now AT&T Archives), University of Illinois archives, the various University of Pennsylvania libraries, and the Open University library. Quotations are the lifeblood of any historical work, and I am indebted to the various institutions who gave me permission to quote material from their collections: University of Akron; AT&T Archives; Beckman Center; B. F. Goodrich & Co.; University of Illinois; National Archives.

Several academic colleagues have encouraged me with their interest, criticism, and advice, including Robert E. Kohler, Sheldon Hochheiser, John K. Smith, Robert Solo, and Robert Friedel. This book was completed at the Open University, Milton Keynes, and I wish to thank my head of department Colin A. Russell for his patience and understanding. I am also grateful to the staff at the University of Pennsylvania Press for its rapid and professional publication of my manuscript.

Peter J. T. Morris
Christmas Day, 1988

Introduction

In the interwar period, natural rubber was a commodity of enormous economic, social, and military importance. Without it, no sophisticated armed forces or advanced economy could operate. Automobiles, which had become a key element of the American social fabric, could not be used without rubber tires. There was a large and technologically advanced American rubber industry, which was nevertheless at the mercy of foreign cartels in peacetime and enemy submarines during a war. As a natural product, rubber was subject to extreme price fluctuations, especially in periods of political uncertainty.[1]

When the Japanese attacked the Dutch East Indies (now Indonesia) and Malaya in the winter of 1941–1942, the United States and its allies were cut off from nine-tenths of the world's rubber-producing regions. A modest stockpile of natural rubber had been built up, more rubber could be recycled, and the search was on for new supplies from countries such as Brazil, but it was clear that the shortfall could be bridged only with synthetic rubber. An industrial miracle was needed to create a synthetic rubber industry before the shortage of rubber severely weakened the war effort.

American expertise in the fields of rubber technology and chemical and petroleum engineering—and the hard work of a legion of scientists and engineers—made the miracle possible. The United States produced nearly 850,000 tons of synthetic rubber in 1945, seven times more than Germany had produced in its peak year of 1943. Both countries made their synthetic rubber from two hydrocarbons, butadiene and styrene. The Germans called it Buna S and the Americans GR-S (Government Rubber-Styrene).

From the outset, it was realized that no modern industrial project on this scale could function properly without the support of a research and development program. Between 1942 and 1956, the U.S. government, through its agent the Reconstruction Finance Corporation, financed a $56 million research and development program. It was given the task of solving production problems and conducting research on synthetic rubber, with the aim of developing a better rubber than the original GR-S. A shortage of qualified manpower in the rubber industry made the participation of university-based researchers vital. During the lifetime of this research program, several advances were made, culminating in the industrial synthesis of natural rubber in 1954.

The rubber research program provides an excellent example of a government-funded cooperative research and development program, geared to the improvement of existing products and the innovation of new ones. The program was characterized by a free exchange of information between universities and industry, and between research groups in different companies. It had been established to meet an urgent national need for rapid scientific and technological advances in the field of synthetic rubber. The rubber research program was funded for a period sufficiently long to demonstrate the success or failure of government-sponsored collaborative research.

There is a striking parallel between rubber in the interwar period and petroleum in the 1970s. The development of an economically viable synthetic rubber reduced America's dependence on external supplies of a crucial material. President Carter used the rubber program as the model for his ill-fated energy program. Recent fears about Japan's growing superiority in high technology, including microchips, computers, and high-performance polymers, have increased demands for government-sponsored programs to encourage increased collaboration between different companies in the same field, and between industry and academe, in the name of "competitiveness."[2] Light-conducting fibers, superconducting materials, and drugs to combat AIDS have also been proposed as key topics requiring rapid scientific and technological innovation by such means.

It is perhaps foolhardy to generalize from the history of one project to other quite different projects or to equate the challenge facing the pharmaceutical and the semiconductor industries today with the situation in the rubber industry forty years ago. However, the cost of failure in technological innovation is often great, in both economic and human terms. It would be even more foolish to overlook the lessons that can be learned from this rare example of a government-sponsored program in an area of scientific and commercial significance. The rubber research program can still offer important indicators to the best way to fund research and promote technological innovation.

To be part of a large crucial program during a world war was an

exciting affair, and scientists who were participants in the rubber research program continue to praise it. They argue that the research program: (1) assisted GR-S production, especially during the key wartime
period; (2) accelerated the development of cold rubber and oil-extended
rubber; (3) aided the innovation of specialty rubbers, particularly for
military applications; and (4) aided the synthesis of natural rubber at
the very end of the program.[3] It is further claimed that the research
program "led to the rapid advances of polymer science that we have
witnessed during the last thirty years,"[4] partly by promoting the social
development of polymer science as a discipline in the 1940s and 1950s.

It is hard to argue with success, and difficult to prove a negative.
Several commentators, however, have been critical of the research program. John T. Cox, a former Deputy Director at the Office of Rubber
Reserve, argued that the synthetic rubber industry in America had "progressed . . . in its technology, but not as far as it would have had it been
in the hands of private enterprise, where the stimulus of . . . competition
would have taken [it] to even greater heights."[5]

Cox was not alone in extolling the virtues of competition. The
Interagency Policy Committee (the Batt Committee), set up by President
Truman at the war's end to formulate government policy on synthetic
rubber, declared in its second report in 1946:

The Committee . . . believes competition in itself to be an extremely powerful
incentive. A research problem involving original thinking and constructive imag-
ination may be better solved by a number of research organizations working
independently and in competition with one another than by one organization
which may have undertaken the problem under Government sponsorship.[6]

Writing in the *Quarterly Journal of Economics* in 1954, economist
Robert Solo delivered his judgment of the program:

In terms of its impact on the actual technology of the synthetic rubber industry,
the government sponsored program appears to have been a failure. The failure
has not been at the level of basic research. There appears to have been a mass
of highly competent research done by individual scientists in companies, institutions and universities. The failure has been to bridge the gap between research
accomplishment and technological innovation. The failure has been in directing
and evaluating research, in selecting and developing the significant discovery.
. . . The program did not fail because it was government-run. It has not been
run by the government—it has been merely paid for by the government.[7]

Twenty-six years later, Solo stated it more bluntly: "The results—
forty million dollars later—were zilch, zero, nothing. By the measure
of technological achievement, the R&D program was useless."[8]

As to what should have happened, Cox and Solo were divided. Cox believed that private enterprise was the proper arena for the development of synthetic rubber and argued that competition would have been more effective than cooperation.

Solo used a more sophisticated argument. The operation of free enterprise methods, he argued, would not work if the outcome desired by the state was not profitable for the companies involved. There were good arguments for a decentralized research program (whereby funds were given to outside bodies) and for a centralized program under direct government control, and both had drawbacks as well. Solo concluded:

But whether the research and development program is centralized or decentralized . . . [it] must be directed and run by a technically competent and responsible authority who stands accountable for the results, and whose position . . . depend[s] directly on the failure or success of that program.[9]

It follows that Solo was not concerned about the lack of competition during the program; indeed he argued the inertia of the program stemmed from the anticipated competition after the program ended.

There are four questions that animate this study. Was the research program a success or a failure as a means of promoting innovation? What does the rubber research program tell us about the relationship between industry and academia, and science and technology? How can government (or corporations) economically and effectively promote innovation? Is free competition a better means of promoting innovation than industry-wide cooperation?

Notes

1. William C. Geer, *The Reign of Rubber* (New York, 1922), is a clearly written and lavishly illustrated account of the early natural rubber industry by a vice-president of B. F. Goodrich Co. Charles Morrow Wilson, *Trees and Test Tubes: The Story of Rubber* (New York, 1943), covers the development of natural and synthetic rubber up to 1942. Howard Wolf and Ralph Wolf, *Rubber: A Story of Glory and Greed* (New York, 1936), contains useful information. For the strategic importance of natural rubber, see Gibson B. Smith, "Rubber for Americans: The Search for an Adequate Supply of Rubber and the Politics of Strategic Materials" (Bryn Mawr College Ph.D. thesis, 1972); and R. F. Chalk, "The United States and the International Struggle for Rubber, 1914–1941" (University of Wisconsin Ph.D. thesis, 1970).

Histories have been written of three of the big four rubber companies. Firestone: Alfred Lief, *The Firestone Story* (New York, 1951). Goodyear: Hugh Allen, *The House of Goodyear: Fifty Years of Men and Industry* (Cleveland, Ohio, 1949); Maurice O'Reilly, *The Goodyear Story* (Elmsford, New York, 1983); M. J. French, "The Emergence of a U.S. Multinational Enterprise: The Goodyear Tire and Rubber Company, 1910–1939," *Economic History Review,* second series, 40 (1987), 64–79. David Dietz's *The Goodyear Research Laboratory* (Goodyear, 1943), written to mark the opening of the new laboratory, is short (57 pages) and disappointingly uninformative. United States Rubber: Glenn D. Babcock, *History of United States Rubber Company* (Bloomington, Indiana, 1966).

The history of several petroleum companies in the rubber program have been recorded. Jersey Standard: Henrietta Larson, Evelyn Knowlton, and Charles Popple, *New Horizons, 1927–1950* (New York, 1971); Popple, *Standard Oil Company (New Jersey) in World War II* (Standard Oil, 1952); Larson and Kenneth Porter, *History of Humble Oil & Refining Company: A Study in Industrial Growth* (New York, 1959). Phillips Petroleum: *Phillips: The First 66 Years* (Phillips Petroleum, 1983). Shell: Kendall Beaton, *Enterprise in Oil: A History of Shell in the United States* (New York, 1959). Also see Peter H. Spitz, *Petrochemicals, The Rise of an Industry* (New York, 1988); and Don Whitehead, *The Dow Story: The History of the Dow Chemical Company* (New York, 1968).

2. James H. Krieger, "Cooperation Key to U.S. Technology Remaining Competitive," *Chemical and Engineering News* (27 April 1987), 24–26.

3. This list is drawn from several sources, chiefly: interview of Paul S. Greer by Peter Morris, for Beckman Center for History of Chemistry (hereafter Beckman Center) oral history program, 13 November 1985; O'Callaghan, *The Government's Rubber Projects,* volume 2, 566–584 (see Chapter 1, note 5 for full citation); Maurice Morton, "History of Synthetic Rubber," in Raymond B. Seymour (ed.), *History of Polymer Science and Technology* (New York and Basel, 1982), 225–239; P. S. Greer, "Office of Synthetic Rubber Research Related to Defense Requirements," in *Proceedings, Joint Army-Navy-Air Force Conference on Elastomer Research and Development* (Washington, D.C., 1954), 36–43.

4. Morton, "History of Synthetic Rubber," 236.

5. J. T. Cox, Jr., *Chemical and Engineering News* 31 (1953), 36.

6. Interagency Policy Committee on Rubber, Second Report, (Washington, D.C., 1946), 41; cited in Robert Solo, *Synthetic Rubber: A Case Study in Technological Development Under Government Direction,* Study No. 18 for the Sub-Committee on Patents, Trademarks and Copyright, Committee on the Judiciary, U.S. Senate, 85th Cong., 2nd sess., 1959, Committee Print, 93; reprinted as *Across the High Technology Threshold: The Case of Synthetic Rubber* (Norwood, Pennsylvania, 1980).

7. Robert Solo, "Research and Development in the Synthetic Rubber Industry," *Quarterly Journal of Economics* 68 (1954), 79.

8. Solo, introduction to *Across the High Technology Threshold,* vii–viii.

9. Solo, "Research and Development," 82.

Chapter 1
The U.S. Government Rubber Program

On the Banks of the Rhine: 1929–1936

The synthetic rubber now used in many automobile tires does not differ significantly in chemical composition from the rubber patented in 1929 by two chemists who worked in the I.G. Farben's Leverkusen laboratories near Cologne.[1] It is remarkable not only that the first combination patented by the German chemical combine—butadiene and styrene—has survived for so long but also that this survival is largely the result of an American research program set up during World War II.

This rosy future would not have been predicted in the early 1930s, even within I.G. Farben. The development of Buna S (as it was then called) was curtailed in 1931 because of the Depression and extremely low natural rubber prices. Apart from being several times more expensive than natural rubber, Buna S was almost impossible to process on existing rubber machinery and it quickly cracked in use. Later that year, however, Du Pont announced that it had developed a synthetic rubber which, in contrast to natural rubber, resisted attack by gasoline, air, and sunlight. Filling demands for gas hoses and gas tank linings, Duprene (later renamed neoprene) soon produced a profit even in the midst of the Depression.

Neoprene's success encouraged I.G. Farben to reassess the results of its earlier research. Buna N, similar to Buna S but with styrene replaced by acrylonitrile, was found to possess the gasoline resistance and other desirable properties that made neoprene profitable. To help solve the processing problems, I.G. Farben sought out a progressive rubber manufacturer. General Tire of Akron, Ohio, agreed to experiment with Buna N on its equipment and the trials were carried out in the spring of 1934. The tests were brought to a premature conclusion when General Tire began to fear that its machinery could be damaged, and the American firm produced a negative report on its experience with Buna N.

The situation changed shortly afterward when the German government demanded that I.G. Farben concentrate its efforts on the development of a tire rubber. Following unsuccessful attempts to obtain a license for neoprene from Du Pont, the German firm revived Buna S, which, unlike Buna N, could be blended with natural rubber. Because of technical problems, labor shortages, and disagreements with the government, I.G. Farben did not achieve a production level of 64,000 long tons a year until the beginning of 1941.

From the Rhine to the Cuyahoga: 1936–1941

This activity did not go unnoticed by the American rubber companies. After they failed to reach an agreement on the terms of a license with I.G. Farben in 1932–1934, Goodrich and Goodyear independently initiated their own research on synthetic rubber. While the two companies were in competition, their research ran along very similar lines and at about the same rate.[2] Indeed, the two firms ordered an identical piece of pilot plant equipment from the same engineering company within weeks of each other in 1936.[3]

Waldo Semon of Goodrich and Lorin Sebrell of Goodyear made trips to Leverkusen in 1937–1938 to exchange information and samples with their counterparts in I.G. Farben. The Germans did not divulge any significant details about Buna S manufacture, and the independent efforts of Goodrich and Goodyear to reach agreement with I.G. Farben were unavailing. The German company could not believe that the two American firms were almost as advanced in this field as it was, and even thought that Goodyear was making its samples (including a complete tire in 1938) from surreptitiously obtained Buna S.

Rather than sign an unequal agreement with I.G. Farben, the American companies spurred their research teams to make a good synthetic rubber that was not covered by the German firm's patents. Indeed, the codename within Goodrich for its first successful rubber was "nirub" for noninfringing rubber. A large number of starting materials (monomers) and rubbers were made by Goodrich and Goodyear, but the attrition rate was high: of the 111 rubbers studied by Goodrich, only 6 were chosen for further study.[4] To obtain a cheap, assured supply of the starting materials, especially butadiene, Goodrich collaborated with Phillips Petroleum, and Goodyear with Shell Development Company and Dow Chemicals.

With the outbreak of war in Europe in 1939, Jersey Standard (who had collaborated with I.G. Farben in the oil-from-coal field) took over the administration of the Buna patents in the United States. The American oil company revived the negotiations with the large rubber companies. The terms proposed were not onerous, but Standard would become the sole supplier of butadiene. This clause made the offer unattractive to Goodrich and Goodyear, who had little to gain from obtaining a license. Goodrich was carrying out experiments to show that Buna S could not be prepared from the patent, as the law requires. On the other hand, Firestone and U.S. Rubber, who had carried out less research on synthetic rubber, were eager to accept.

Goodrich displayed its contempt for Jersey Standard's attempts to coordinate the development of synthetic rubber by publicly announcing on 5 June 1940 the launch of Ameripol (American polymer) synthetic rubber in which methyl methacrylate replaced styrene. Patriotic citizens and companies were supplied with Liberty tires that combined Ameripol with natural rubber—at a 30% premium over the normal price for tires—during July. Goodrich now had a plant that could produce 2,000 long tons of synthetic rubber a year. By contrast, Goodyear kept a low profile. Its synthetic rubber Chemigum, which was similar to Buna N, had been available in experimental quantities since 1938 and the rubber company was assisting potential customers with their trials. In November 1940, Goodyear started up a ton-a-day Chemigum plant.

After waiting for over a year, Jersey Standard issued writs against Goodrich and Goodyear for infringement of its patents in September 1941. This dispute was resolved three months later when a patent and information sharing agreement with the Rubber Reserve Company— the "December 19th agreement"—was signed by Standard and the four rubber companies.

The Emergency Program: 1940–1942

The Rubber Reserve Company (RRC) had been formed in June 1940, when President Roosevelt declared rubber to be a strategic and crucial material. U-boat attacks on Atlantic shipping, and the fear that the Japanese might similarly hinder access to the rubber-producing areas in Asia, had increased official concern about the rubber supply. The company's main function was the stockpiling of natural rubber, but the RRC took over the functions of the Francis Committee in October 1940. This committee, chaired by Clarence Francis, president of General Foods, had been set up in the same month as the RRC by the Advisory Committee of the Council for National Defense to establish a synthetic rubber industry. The Francis Committee had developed a plan for 100,000 long tons/year program with four plants, located in the private sector, but with financial inducements from the Federal Government.[5]

Between the fall of 1940 and the summer of 1942, the government's synthetic rubber policy was set by Jesse Jones, a Texas banker and conservative Democrat, who was Roosevelt's business policy advisor, Federal Loan Administrator, Secretary of the Commerce Department, and first chairman of the RRC. Jones decided that the 100,000 long tons/year target was excessive and scaled it down to 40,000 long tons/year. By the spring of 1941, however, the stockpiling of natural rubber was progressing—over 30,000 long tons had arrived from Indochina before it was occupied by the Japanese—and Jones was confident that a stockpile sufficient for three years' consumption was within reach. In March 1941, the four rubber companies and the RRC therefore agreed to a new program, whereby the four 10,000 long tons/year plants would be erected but fitted out with only enough equipment to produce a quarter of that output. Under pressure from the Office of Production Management, the RRC reinstated the full 40,000 long tons/year program on 16 May and concluded the necessary contracts on that basis.

Following the Japanese air raid on the U.S. naval base at Pearl Harbor on 7 December 1941, the RRC increased the planned capacity to 30,000 long tons/year in each plant; technical developments had already raised capacity from 10,000 to 15,000 long tons/year. During January 1942, Jones expanded the program still further to 400,000 long tons/year. In the spring of 1942, as the situation in the Pacific and the Far East continued to deteriorate, a program for the production of 805,000 long tons/year of synthetic rubber was developed by the RRC and the War Production Board (WPB), the successor of the Office of Production Management. Of the 705,000 long tons/year of GR-S (the American version of Buna S) planned, priority was to be given to the production of 350,000 long tons in 1943.

Donald Nelson of the WPB appointed Arthur Newhall his rubber coordinator, but the autocratic Nelson did not give his subordinate any significant authority. Much of the burden of providing scientific advice was carried by Edward Weidlein, director of the Mellon Institute, but he, too, lacked the authority to settle the technical disputes that arose in this period.

Controversy on the Potomac

The rubber program became controversial during the summer of 1942, despite the progress made since Pearl Harbor. The Justice Department brought an anti-trust action against Jersey Standard, alleging a conspiracy with I.G. Farben that had weakened America's ability to defend herself. This exposed an already unpopular company to even more public criticism. The rubber situation was becoming critical: stocks were running low as the three-year stockpile had not materialized. President Roosevelt was reluctant to order rubber rationing or a speed limit to save rubber and gasoline because of the Congressional elections in November.[6]

In this heated atmosphere, Guy Gillette, a populist senator from the farm state of Iowa, criticized the RRC's decision to obtain most of the butadiene needed from petroleum (and hence Standard Oil) and not grain alcohol. A significant amount of butadiene was to be produced from alcohol by Union Carbide, but this was synthetic alcohol derived from petroleum, not natural alcohol. The RRC had originally decided against natural alcohol because the usual raw material for industrial alcohol, Cuban molasses, was placed in doubt by U-boat attacks and the heavy demand for shipping elsewhere. Furthermore, alcohol was sorely needed for explosives manufacture.

When it became clear in early 1942 that there would be a large grain surplus (because of Federal farm subsidies), and that the whiskey industry's facilities would be available for the conversion of this grain into alcohol, the RRC reconsidered its position. At this point, the Gillette Senate committee opened its hearings. The WPB and RRC officials, caught in the middle of their reassessment of the various sources of butadiene, appeared hesitant, even ignorant, about the possible options. On a wave of patriotism and anti-Standard Oil feeling, Congress passed Gillette's bill at the end of July 1942. With an eye on farm belt votes, it compelled the RRC to use grain alcohol, and created a "rubber czar" who would be independent of the executive branch of government.

President Roosevelt vetoed the bill on 6 August but, acknowledging the public's concern about the rubber situation, he appointed a three-man commission to look into this issue. The committee was chaired by the financier and power broker, Bernard Baruch. The other two members were university presidents and scientists: physicist Karl T. Compton of the Massachusetts Institute of Technology and chemist James B. Conant of Harvard.

The committee submitted its report barely a month later, on 10 September 1942. Among its recommendations, which were accepted by the administration, were the introduction of a maximum speed limit (of 35 mph!) to conserve rubber, the expansion of rubber reclaiming facilities, and the appointment of a Rubber Director to push the program through. The committee also increased the amount of butadiene to be produced from grain alcohol and recommended the adoption of a "quick" butadiene program to obtain more butadiene from oil refinery gases. Ironically, these two proposals were later pared back or abandoned by the Rubber Director. The committee also called for an increase in GR-S production from 705,000 to 845,000 long tons/year.

Research Program Established

Following publication of the Baruch Committee's report, Roosevelt appointed William Jeffers, president of the Union Pacific Railroad, as the first Rubber Director.[7] Unfortunately, the Baruch Committee had (perhaps deliberately) left the nature of the relationship between the Rubber Director and the chairman of the WPB unclear. There was considerable friction between Jeffers and Nelson over the priority to be given to the rubber program. In a stormy letter to Nelson at the end of 1942, Jeffers roared:

As my Chairman, you predict failure for the rubber project as I am administering it and as I propose to administer it. . . . [Your letter] casts a degrading eclipse over my organization . . . you have no confidence in my ability to solve the rubber problem and you should be relieved of this responsibility, either by my immediate retirement or by permitting me to report directly to the President. . . . There can be no compromise. Either you are right or I am.[8]

Partly because of these disagreements, Jeffers returned to Union Pacific in September 1943.

Jeffers was replaced by the Deputy Rubber Director, Colonel Bradley Dewey, a chemical engineer. A year later, Dewey returned to his own firm of Dewey and Almy and the position of Rubber Director was abolished. The Rubber Reserve Company, which had continued to administer routine production matters, assumed most of the Rubber Director's functions. The RRC was replaced in June 1945 by the Office of Rubber Reserve (ORR) (later the Office of Synthetic Rubber) within the Reconstruction Finance Corporation (RFC). President Herbert Hoover had founded the RFC in 1932 to lend money to ailing banks. During the New Deal period, this facility was extended to industry and agriculture.

Shortly after his appointment as Deputy Rubber Director, Dewey invited Robert R. Williams, leader of the chemical section of Bell Telephone Laboratories to lead the research effort in October 1942.[9] Dewey then asked Bell Laboratories for the loan of Calvin S. Fuller, the leader of its polymer research, to assist Williams. R. R. Williams held several attractions for Dewey. Like Dewey, Williams had served in the Chemical Warfare Service in 1918.[10] Even before Bell Telephone Laboratories was founded in 1925, he had established a major program of cable insulation and rubber research at its predecessor, Western Electric Engineering. Williams was an expert on natural rubber and knew as much about polymers as any research director in America. He was renowned for his isolation and synthesis of thiamine (vitamin B1), which aided the conquest of the tropical disease beriberi.

Williams and Fuller spent the early months of the research program, up to the end of 1942, visiting laboratories, selecting academic research groups to work in the program, and clarifying the problems to be investigated.[11] In December, eleven universities were invited to join the rubber research program. The key figures were Carl (Speed) Marvel at the University of Illinois, Izaak (Piet) Kolthoff at the University of Minnesota, and W. D. Harkins at the University of Chicago. Williams and Fuller appear to have been unaware of the importance of Morris Kharasch's work at the University of Chicago, but his group was added to the program a few weeks later at the suggestion of Harkins and Marvel.

The process of setting the program's aims was overshadowed by anxiety, confusion, and ignorance. Much of the early confusion centered on the correct balance of short-term trouble-shooting and fundamental long-term research. Williams described the position at the beginning of the research program:

Research is going forward with all possible pressure for results of early applicability but with a recognition that proper large scale confirmation of new results will probably be delayed beyond the most crucial period with respect to rubber supply. There is no escape from this probability and the research program should be regarded primarily as costly insurance against still more costly future delays due to unknown defects in the present process and plant.[12]

R. R. Williams resigned in May 1943, and Calvin Fuller became head of the research branch. Fuller returned to Bell Laboratories on 1 June 1944 and his place was taken by Raymond Dunbrook, who had worked for Firestone. After the research administration reverted to the Rubber Reserve Company in September 1944, the Research and Development Branch was directed by Evan Boss, who had been responsible for compounding research in the Office of the Rubber Director, but Dunbrook remained in charge of research.

Creation of a New Industry: 1942–1945

The synthetic rubber plants came into production relatively smoothly and, with one exception, reached or exceeded their rated capacity. The four plants in the original program started up during 1942, and all fifteen plants were in operation by the end of November 1943. The first butadiene-from-alcohol plant came on stream in January 1943; the other two started up in the summer. They were soon operating at up to twice their rated capacity.[13]

By contrast, the butadiene-from-petroleum plants labored under a shortage of crucial materials needed to construct the plant, competition from the aviation fuel program for the intermediate butylenes, and various technical problems. Although 184,000 long tons of petroleum-based butadiene capacity should have been available in 1943, only 27,750 long tons were produced.[14] While the alcohol plants also fell short of their 1943 target, they nevertheless produced 157,435 long tons, and continued to carry the burden during 1944.

This butadiene shortage led to a shortfall of synthetic rubber in 1943 and early 1944, but it is uncertain that the rubber processing plants could have handled the planned totals. GR-S production rose from a tiny 3,721 long tons in 1942 to 182,259 long tons in 1943, and then increased threefold to 670,268 long tons in 1944. By early 1945, with the problems in the petroleum-based plants solved, GR-S production was running ahead of demand, and production at the more expensive alcohol-based plants was cut back. Total GR-S production in 1945 was 756,042 long tons. Fortunately for the rubber program, and the Allied war effort, the peak military demand for rubber did not occur in 1943, as predicted, but in the summer of 1944.[15]

The wartime research—especially within the rubber industry—was concerned with the incremental improvement of existing processes. If the polymerization was allowed to take place unchecked, the rubber molecules were excessively long, eventually forming a gel. The over-long and branched molecules made the rubber difficult to process and the gel had a bad effect on the physical properties of the rubber. These undesirable processes could be reduced, but not prevented, by stopping the polymerization when only 72% of the monomers had reacted (72% conversion), and by adding a small amount of a mercaptan as a "modifier." These two steps had been introduced into American practice before Pearl Harbor. The mercaptan modifiers had been patented by I.G. Farben in 1937, but the Germans did not use them to make Buna S until 1943.[16]

At the beginning of the GR-S production program, there were puzzling discrepancies in the length of time that elapsed before the polymerization began (induction period). A collaborative effort between the soap industry (especially Procter and Gamble), the rubber industry, and the University of Illinois soon traced the variation to the presence of certain fatty acids in the soap. The soap also hindered the recovery of the unreacted monomers, because of excessive foaming during the steam-distillation of styrene. This led to the introduction of anti-foaming agents such as candelilla wax and silicones.[17] Another problem during the recovery process was the formation of "popcorn" polymer by butadiene. Morris Kharasch showed this polymerization could be prevented by adding a small amount of sodium nitrite (which also gives ham its characteristic red color).[18]

The original modifier used in the rubber program was "Lorol," a mixture of primary mercaptans made from coconut oil. From the outset, it was known that the modifier was consumed during the polymerization, typically vanishing around 40% conversion. This meant that a soupy "overmodified" rubber was produced at the beginning of the polymerization, and the hard over-long "undermodified" rubber once the modifier had disappeared. This mixture was not altogether undesirable; the soupy material plasticized the hard stuff that gave the rubber much of its strength. However, it presented problems during the curing process, and was blamed for poor processibility. Considerable effort was therefore made to produce a "controlled" or constant viscosity rubber that was presented with the same amount of modifier throughout the polymerization. Research by the rubber companies and three universities (Harkins at Chicago, Kolthoff at Minnesota, and William Reynolds at Cincinnati) had shown by 1945 that replacement of the original primary mercaptans by tertiary mercaptans made by Phillips Petroleum from refinery gases was the best way of preventing the premature disappearance of the modifier. Even before the results of the academic studies were in, interest had increased in these mercaptans because of the growing shortage of coconut oil.

The properties of GR-S were connected to the amount of styrene present in the polymer, and it was important that the butadiene-styrene ratio be kept constant. As the proportion of styrene to butadiene in GR-S bore no direct relationship to the ratio of the two monomers in the reactor, it was necessary to analyze the rubber for its styrene content. William O. Baker at Bell Telephone Laboratories developed a method that derived the styrene content from measurements of the refractive index of GR-S solutions. A cheaper and simpler method, which measured the refractive index of the solid rubber, was introduced in 1944 by the National Bureau of Standards.

Chemical analysis played a key role in the wartime research program. It was essential not only for quality control but also for the research on the mechanism of GR-S polymerization and the chemical structure of the rubber. New analytical techniques were developed by Piet Kolthoff at the University of Minnesota and his former student Herbert Laitinen at the University of Illinois. As a result of his analytical work, Kolthoff was drawn into the investigation of how the various parameters—temperature, soap, acidity, and modifier—influenced the rate or the outcome of the polymerization. Kolthoff's colleague Edward

Meehan studied the variation in the styrene content of GR-S during the course of polymerization using the then novel ultra-violet spectrophotometer. William Draper Harkins at the University of Chicago developed new techniques for the study of soap emulsions. In 1946, he published a new model for these emulsions. He located the initial site of polymerization in tiny droplets of monomer in the aqueous phase, which were then converted into polymer particles soaked with fresh monomer. At Cornell University, the emigré Dutch physicist Peter Debye developed mathematical equations that permitted the determination of the molecular weight of GR-S and other polymers by measuring the visible light scattered by the polymer in solution.

The Postwar Period: 1945–1950

Because of the threat posed by the Soviet Union, it was necessary on the grounds of national security to maintain a significant synthetic rubber industry after the war ended.[19] Synthetic rubber production was naturally cut back, but it was nevertheless maintained at 300,000 to 400,000 long tons of GR-S a year. For the same reason, the plants closed down were mothballed rather than scrapped, or sold on the condition that they could be reactivated for synthetic rubber production at short notice. Government control of the plants and rubber inventories was retained, but the Rubber Act of 1948 called for the disposal of the plants by 1950.

When Boss resigned as head of the Research and Development Branch at the beginning of 1946, a most unusual arrangement was introduced. The Research and Development Branch was headed by the research leaders of the four major rubber companies (Harlan Trumbull of Goodrich, Morris Shepard of U.S. Rubber, John Street of Firestone, and Albert Clifford of Goodyear) in rotation, four months at a time.[20] Dunbrook was succeeded as research section chief by Alvin Borders from Goodyear in June 1946. Borders left after only four months and was replaced by James D'Ianni, also from Goodyear.

The rotating leadership ended in August 1947, and Oliver Burke was appointed chief of the Research and Development Division. Paul Greer and Thomas Swan, who had replaced D'Ianni, acted as his assistants. Burke left in October 1950 to set up a consulting firm connected with the development of synthetic rubber and was replaced by Greer, who had slowly worked his way up the administrative ladder.

This was a golden period for the research program. The research groups were now free to pursue long-term goals, and research funds continued to flow at an annual rate of $3.5 million, despite the drop in synthetic rubber production. The main aim of the immediate postwar research was the development of a rubber that could compete with natural rubber on both quality and price.

In October 1948, the Reconstruction Finance Corporation announced the introduction of "cold rubber," a superior form of GR-S, which was an excellent rubber for automobile tire treads. This new rubber was made possible by a new catalyst system which permitted the polymerization of butadiene and styrene at 5°C (41°F) instead of the standard 50°C (122°F), hence the name "cold" rubber. The drop in polymerization temperature produced a rubber that was strong but without the harmful gel found in "hot" GR-S. However, cold rubber was not simply the old GR-S with a lower polymerization temperature. Its manufacture incorporated other advances, such as the tertiary mercaptans mentioned above, a new kind of soap for the emulsion, and greatly improved carbon blacks (for tire reinforcement) developed by Phillips Petroleum. With improvements in monomer recovery, the polymerization could now be stopped at 60% conversion instead of 72%.

Cold rubber was the end result of a decade's research by the four large rubber companies and Phillips Petroleum. The universities had assisted this development with their studies of the underlying processes. Piet Kolthoff played a notable role in the introduction of the new modifiers and soaps, and in the development of the new catalyst system. Nevertheless, the "Custom" recipe adopted by Rubber Reserve in 1948 for the standard production of cold rubber was based largely on research by Williams Reynolds and Charles Fryling at Phillips Petroleum.

In 1950, cold rubber already accounted for 38% of all GR-S production. It was the equal of natural rubber for automobile tires, but was also difficult to process and no cheaper than the old GR-S.

The Korean War: 1950–1953

After the Communist invasion of South Korea in June 1950, the focus of the program switched from quality to quantity. Even before the war began, the price of natural rubber had been spiralling upward, partly because of heavy buying by the Soviet Union. It reached a peak of 73 cents/lb compared with the fixed price of 18.5 cents/lb for GR-S. By September, the synthetic rubber industry had been completely reactivated. GR-S production in 1951 was 697,000 long tons, almost double the previous year's total. To cover the additional cost of the alcohol-based butadiene from the standby plants, the government raised the price of GR-S to 24.5 cents/lb.

Oil-extended rubber—a 4:1 mixture of cold rubber and selected mineral oils—was introduced independently in 1951 by General Tire and Goodyear as a way of increasing the volume of rubber production.

The addition of mineral oil also made cold rubber easier to process and cheaper. The output of GR-S could be increased in other ways. The Government Laboratories in Akron (founded in 1944) and Goodrich collaborated on the development of a "rapid" GR-S, which used a catalyst system created for cold rubber to reduce the polymerization period at 50°C. This doubled the rate of production and hence greatly increased plant capacity.

While cold rubber was adequate for automobile tires, it was not suitable for an all-synthetic large truck tire (or the similar aircraft tires). Truck tires usually operate at high temperatures under heavy loads for long periods. GR-S tires suffered excessive heat build-up while running, and this problem was accentuated by the synthetic rubber's poor strength temperatures. Goodrich had hoped to solve this problem during 1942, but it remained refractory.

In March 1950, Goodrich converted part of its pioneering research section (unconnected with the government program) into the "American Rubber Team," directed by Waldo Semon, with a mission to develop a synthetic rubber that could be used to make heavy-duty tires. The group—fifty strong at its peak—failed in its main objective, but several advances in rubber technology were made.[21]

Carl Marvel's group at the University of Illinois attempted to solve this problem by replacing styrene with compounds related to benzala-cetophenone, but no significant improvement in performance was obtained, given the higher cost of the new polymers. This project initiated Marvel's research on heat-resistant fibers in the 1960s and 1970s, which resulted in the commercialization of polybenzimidazole (PBI) fiber in 1983.[22]

Another long-standing problem was the search for a synthetic rubber that could withstand very low temperatures, after GR-S tires "froze" solid during military maneuvers in the far north in 1946. This led to the setting up of the arctic rubber program, which was funded by the Department of Defense rather than the Reconstruction Finance Corporation. At a conference of the leading chemists in the research program in January 1949, chaired by Charles Price of Notre Dame University, various exotic rubbers were proposed including fluorocarbon and silicone polymers.[23]

However, the general-purpose arctic rubber adopted in 1950 was a GR-S rubber (hot or cold) with less styrene—10% or 15% instead of 25%—than the standard GR-S.[24] This was more or less the same as Buna S10, developed by I.G. Farben during World War II for use on the Russian front. The arctic rubber program then sought a fuel-resistant rubber with better low-temperature properties than Buna N. The fluorocarbon rubbers were considered the most promising. The silicone rubber were unsuitable because they were attacked by fuel oil.[25]

The Disposal of the Plants: 1953–1955

Even when the Korean War faded into the background in 1952 and 1953, GR-S production remained high at an annual level of 600,000 long tons. The superior properties of cold rubber and the cheapness of oil-extended rubber (31% of cold rubber and 15% of all GR-S in 1953) made GR-S more competitive with natural rubber. This lent increased impetus to attempts to put the synthetic rubber industry in the private sector. The 1950 deadline in the 1948 Rubber Act had been extended to 1952, and then to 1954, because of the war.

The Reconstruction Finance Corporation (RFC) presented its disposal plan to Congress in 1953.[26] The RFC and Congress were anxious to prevent the big four rubber companies from forming an oligarchy in synthetic rubber production, and to ensure that the plants would be available in a national emergency for GR-S production. Congress rejected the RFC's plea that it be put in charge of the disposal operation, and the 1953 Rubber Producing Facilities Disposal Act created an independent commission for this purpose. The Attorney General was instructed to monitor the disposal process to prevent violations of the anti-trust laws. The twenty-nine plants were to be widely advertised for sale, and the sale contracts would contain a clause that the plants would be available for emergency use for ten years after the sale.

The disposal commission, headed by Leslie Rounds, a former vice-president of the Federal Reserve Bank, met for the first time on November 1953, and the advertisements appeared eight days later. The deadline for bids was 27 May 1954, and by that date seventy-five proposals had been received from thirty-five bidders. The number of bids per plant ranged from eleven for the Los Angeles styrene plant to none at all for the Institute, West Virginia, GR-S plant and the two butadiene-from-alcohol plants. The commission then negotiated with the bidders (the details of the bids were kept secret) to raise the prices bid and to make them more even across the board. It was determined not to sell a plant unless a fair price could be obtained. The first contract was signed on 16 December and the final one on 27 December, the last day permitted by the Act. Twenty-four of the twenty-nine plants had been sold; the bid for the Baytown, Texas, GR-S plant was rejected as too low. The sale of the plants, not including the inventories, raised just over $260 million. Congress approved the sales in March 1955, and the transfers took place in late April. The Baytown plant was handed over to United Carbon on 15 July 1955, following a second round of bids, and the U.S. government was out of the synthetic rubber business.

This period also witnessed a major turnaround in synthetic rubber research. The problem of the large truck tire was finally solved by synthesizing natural rubber itself. Firestone started a special research effort in 1952, determined to succeed where Goodrich's American Rubber Team had failed. Its research director, Frederick Stavely, decided to concentrate on the polymerization of isoprene (the monomer of natural rubber) with the active metal, lithium. This line of research succeeded in making a polyisoprene rubber in early 1953, and by September a pilot plant was in operation. Polyisoprene can have three arrangements in three-dimensional space: all-*trans,* all-*cis,* or a mixture of the two. Natural rubber owes its superior properties to its all-*cis* arrangement. Firestone's original polyisoprene had a lower *cis* content, but Stavely's team was soon able to produce "coral rubber" with a high *cis* content.

Unaware of Firestone's success, Goodrich heard of a new catalyst system, which had been developed in West Germany by Karl Ziegler to polymerize ethylene. Goodrich-Gulf (a joint subsidiary of Goodrich and Gulf Petroleum) obtained a license to use this catalyst from Ziegler in the summer of 1954. A young chemist, Samuel Horne, was detailed to make a copolymer of ethylene and isoprene, and unwittingly prepared an analogue of natural rubber. A public announcement of the new process was made by Goodrich-Gulf on 2 December 1954. Commercial production of Ameripol SN (Synthetic Natural) began in 1958.

Within a year, three other companies had independently developed processes for the manufacture of synthetic "natural rubber" and the similar all-*cis* polybutadiene. At Phillips Petroleum, Robert Zelinski prepared an all-*cis* polyisoprene in December 1954, only to learn of Horne's work. He went on to develop all-*cis* polybutadiene, which was commercialized by Phillips in 1959. The chemists at Goodyear, who did not know the details of the Goodrich process, also succeeded in producing an analogue of natural rubber with a Ziegler catalyst in the summer of 1955. Goodyear started production of Budene (*cis*-polybutadiene) at the end of 1961 and Natsyn synthetic "natural rubber" followed four months later.[27] Finally, Lee Porter of Shell Chemicals used a lithium-cobalt catalyst to produce a polymer like Firestone's coral rubber. "Shell Isoprene Rubber" also went into industrial production in 1959.

The End of the Research Program: 1955–1957

While the academic groups made significant contributions to the scientific basis of the synthetic rubber industry, they made a much smaller contribution in terms of technological innovation. Carl Marvel at the University of Illinois had developed a rubber that was produced by the action of sodium metal on butadiene, or butadiene and styrene. This was superior in some respects to GR- S, but it was never commercialized. This was also true of the Alfin rubbers developed by Avery Morton at the Massachusetts Institute of Technology. The two types of rubber failed for the same reasons. The methods of polymerization were much more expensive and troublesome than the emulsion process. Furthermore, these rubbers could not be easily "modified," and the very long polymers formed were difficult to process.

Marvel's group also attempted to replace styrene with other monomers in the hope of obtaining superior characteristics for a number of end-uses, for example, arctic rubber. Ten years and 150 new monomers later, Marvel was forced to admit defeat: styrene and acrylonitrile were still the best monomers to use with butadiene. The failure of the universities to contribute materially to the innovation of the various specialty rubbers that were developed in the early 1950s—this failure was particularly prominent in the case of synthetic "natural rubber"—doubtlessly contributed to the demise of the research program.

The other surviving element of the research program, the Government Laboratories in Akron, also failed to present a convincing case for continued federal support. Founded in 1944 as an evaluation laboratory and pilot plant, the Government Laboratories assisted with the development of cold rubber and evaluated a large number of new polymers, including many produced by Marvel's group. The close control exercised by the rubber industry over the research program naturally inhibited the development of new rubbers in the Government Laboratories in competition with the large rubber companies. Not needed by these firms as an independent evaluation laboratory, and unable to find a niche as a contract research establishment, the sale of the Akron laboratories to its neighbor Firestone was almost inevitable.

It was argued that the federal government should support further research on rubbers that were of no commercial value but might be needed for national security purposes.[28] Since the late 1940s, however, there had been a shift in funding for such projects from the Reconstruction Finance Corporation to the military research offices. By 1954, the armed forces were sponsoring a significant (possibly even a major) proportion of the academic research on specialty rubbers. The U.S. Air Force, for example, was interested in heat-resistant rubbers for jet-engine O-ring gaskets.[29] In the late 1950s, Marvel also studied the manufacture of synthetic rubbers and other polymers from unsaturated chemicals derived from agriculture with a grant from the U.S. Department of Agriculture, but no commercially viable polymers were found.[30]

The 1953 Disposal Act made provision for the continuation of the research program for a year after the sale of the production facilities, at which point a final decision would be made. In May 1955, it was decided that the government-sponsored research would continue for at least another year under the supervision of the National Science Foundation. This program was allowed to lapse at the end of 1956, and the Government Laboratories were sold to Firestone.[31]

Since the Program: 1955–1985

Total SBR (styrene-butadiene rubber: GR-S) capacity in the United States expanded from 890,000 long tons in 1955 to 1.4 million long tons in 1957. The Goodyear plant in Houston, Texas, alone had a capacity of 200,000 long tons in 1957.[32] American SBR production (excluding latex) topped 1.5 million long tons in 1973, but the industry then went into a decline because the oil crisis made synthetic rubber more expensive relative to natural rubber.[33] Except for a brief rally in 1977–1979, when production almost touched 1.4 million long tons, U.S. SBR production has hovered around 900,000 long tons.[34] Goodyear and Firestone have maintained their position as the two dominant firms in the American industry. Goodrich was in third position for many years, but it was overtaken by Texas-US Chemicals (a joint subsidiary of Texaco and Uniroyal, as U.S. Rubber was renamed in 1967) in the 1970s.[35]

Since the government plants were sold in 1955, there has been only one major technological innovation in the synthetic rubber field: the ethylene-propylene copolymers and terpolymers. This completely new class of synthetic rubber was made possible by Ziegler's new catalyst system, which was improved in the mid-1950s by Giulio Natta of Milan Polytechnic. Whereas butadiene can be polymerized to a rubbery material on its own, neither ethylene or propylene can be made into rubbers by themselves. The first ethylene-propylene rubber was commercialized by Jersey Standard in 1960, and independently by the Italian chemical company Montecatini in 1961. The original copolymer could not, however, be easily vulcanized. The superior terpolymer, with a small amount of a third monomer that permits cross-linking, was announced by Du Pont in 1961 and a semi-industrial plant started up in July of that year. Dunlop Canada introduced their own terpolymer, with a different termonomer, in 1962. Du Pont opened its full-scale plant in 1963, but Dunlop Canada and its partner, Hercules, never got beyond the pilot stage. By 1967, Uniroyal, Copolymer Corporation, and Jersey Standard were producing ethylene-propylene rubber using a termonomer patented by Du Pont.[36]

The hallmark of the post-1955 industry has been technology transfer rather than innovation. The major West German synthetic rubber producer Chemische Werke Hüls (a former joint subsidiary of I.G. Farben and a coal firm) decided to abandon its homegrown but obsolescent technology and purchase a GR-S plant from Firestone in 1954. Two years later it bought the technology from the Houdry Process

Corporation to make butadiene from petroleum.[37] Synthetic rubber production began in Britain and Italy in 1957, in France and Japan two years later, and in Brazil—the original home of natural rubber—in 1962. By the early 1970s, Japan had established herself as a major producer of not only SBR but also all-*cis* polybutadiene and polyisoprene. However, the American industry has retained its dominance: Japanese production of SBR is only about half the level of American production.[38]

Notes

1. German Patent, 570,980, filed 21 July 1929. For the history of the German synthetic rubber industry, see Peter J. T. Morris, "The Development of Acetylene Chemistry and Synthetic Rubber by I.G. Farbenindustrie Aktiengesellschaft: 1926–1945" (Oxford University D.Phil. thesis, 1982). Erich Konrad, "Über die Entwicklung des synthetischen Kautschuks in Deutschland," *Angewandte Chemie* 62 (1950), 491–496. Franz I. Wunsch, "Das Werk Hüls. Geschichte der Chemische Werke Hüls AG in Marl von 1938 bis 1949," *Tradition* 9 (1964), 70–79. Claus Heuck, "Ein Beitrag zur Geschichte der Kautschuk-Synthese: Buna-Kautschuk I.G. (1926–1945)," *Chemiker-Zeitung* 94 (1970), 147–157. Paul Kränzlein, *Chemie im Revier—hüls* (Düsseldorf and Vienna, 1980). Heino Logemann and Gottfried Pampus, "Buna S—Seine grosstechnische Herstellung und seine Weiterentwicklung—ein geschichtlicher Überblick," *Kautschuk und Gummi, Kunststoffe* 23 (1970), 479–486. For the history of neoprene, see John K. Smith, "The Ten Year Invention: Neoprene and Du Pont Research, 1930–1939," *Technology and Culture* 26 (1985), 34–55.

2. For general accounts of the prewar work on synthetic rubber in the United States, see Frank A. Howard, *Buna Rubber: The Birth of an Industry* (New York, 1947), 35–76. Davis R. B. Ross, "Patents and Bureaucrats: U.S. Synthetic Rubber Developments Before Pearl Harbor," in Joseph R. Frese, S.J., and Jacob Judd (eds.), *Business and Government* (Tarrytown, New York, 1985), 119–155. Samuel E. Horne, "The History of Synthetic Rubber," in *Rubber Division 75th Anniversary* (Akron, 1984), 5–7. For Goodrich, see Waldo L. Semon, "Thirty Years' Contributions to the Science of Synthetic Rubber," *Chemical and Engineering News* 21 (1943), 1613–1619; and ACS Rubber Division interview of Waldo Semon, 16 January 1966, stored in University of Akron archives. For Goodyear, see the interview of E. Ralph Rowzee and Ernest J. Buckler by Peter Morris, 26 May 1986. James D. D'Ianni, "Fun and Frustrations with Synthetic Rubber," *Rubber Chemistry and Technology* 50 (1977), G67–68. Hugh Allen, *The House of Goodyear* (Cleveland, Ohio, 1949), 436–448. For Ostromislensky's early research at U.S. Rubber, see Glenn Babcock, *History of the United States Rubber Company,* 244–246.

3. Rowzee and Buckler interview.

4. Semon interview.

5. The most complete history of the synthetic rubber program is the unpublished history of the government's rubber projects compiled by the Reconstruction Finance Corporation in 1948. The first volume, which covered the natural rubber projects, was never revised. The second volume, which recorded the history of the synthetic rubber program (and the scrap rubber program), was revised twice. The 1948 history was written by Brendan J. O'Callaghan. The revisions were carried out by Dorothy Rigdon in 1954 and "under the supervision of Bertram H. Widmer" in 1955. The original typescript, entitled "The Government's Rubber Projects (Vol. II)," is stored at the National Archives

in Washington, D.C., Record Group 234, Office of the Secretary, Entry 26, PI–173. A photocopy is available at the Beckman Center. For convenience, we will cite the history of the synthetic rubber program as O'Callaghan, *The Government's Rubber Projects,* volume 2. For the period covered in this section, see pages 355–429. Also see Howard, *Buna Rubber,* 77–187, and Vernon Herbert and Attilio Bisio, *Synthetic Rubber: A Project That Had to Succeed* (Westport, Connecticut, 1985), 39–68. For the importance of reclaimed rubber in the early war years, see Williams Haynes and Ernst A. Hauser, *Rationed Rubber and What To Do About It* (New York, 1942); Howard Wolf, *The Story of Scrap Rubber* (Akron, Ohio, 1943); and John M. Ball, *Reclaimed Rubber: The Story of An American Raw Material* (New York, 1947).

6. O'Callaghan, *The Government's Rubber Projects,* volume 2, 430–452. Howard, *Buna Rubber,* 188–222. Herbert and Bisio, *Synthetic Rubber,* 68–74. William M. Tuttle, Jr., "The Birth of an Industry: The Synthetic Rubber 'Mess' in World War II," *Technology and Culture* 22 (1981), 35–67. Robert Solo, *Synthetic Rubber,* 1–82. "Rubber—How Do We Stand?" *Fortune* 25 (June 1942), 94–96, 192–194. James B. Conant, Karl T. Compton, and Bernard M. Baruch, *Report of the Rubber Survey Committee* (Washington, D.C., 1942).

7. C. S. Fuller, "History of the Polymer Research Branch, Period December 2, 1942 to June 1, 1944," memorandum presented to the 17th meeting of the Rubber Research Board, 14 June 1944, National Archives, Washington, D.C., Record Group 234, Rubber Reserve Company (hereafter RG 234, RRC), Entry 231, PI–173, Minutes of the Rubber Research Board, typescript, 11 pp. Interview of C. S. Fuller by James J. Bohning for the Beckman Center oral history program, 29 April 1986.

8. Letter from W. M. Jeffers to D. M. Nelson, chairman of the War Production Board, 9 December 1942, RG 179, Records of the War Production Board, Entry 1, Box 205, WPB 036.004, Rubber Director's office, WPB-Administration, PI–15. This letter was written in reply to a letter by Nelson, 8 December.

9. Bradley Dewey to O. E. Buckley, President of Bell Telephone Laboratories, 6 November 1942, requesting the loan of R. R. Williams; Dewey to Buckley, 10 November 1942, confirming the Office of Rubber Director's telephonic request for C. S. Fuller; Buckley to Dewey, 16 November 1942, agreeing to these requests. These letters are in Box 77, Oliver E. Buckley Papers, AT&T Archives, Warren, New Jersey.

10. See the biography of Robert R. Williams in *National Cyclopedia of American Biography* 58 (1979), 411–413.

11. C. S. Fuller and R. R. Williams, "Summary of Present Research on Synthetic Rubber," report to the first meeting of the Rubber Research Board on 6 January 1943 (but clearly written at the end of November 1942), RG 234, RRC, Entry 231, PI–173, Minutes of the Rubber Research Board, typescript, 18 pp.

12. R. R. Williams, "Report of Polymer Research Section," report to the first meeting of the Rubber Research Board on 6 January 1943, RG 234, RRC, Entry 231, PI–173, Minutes of the Rubber Research Board, 3.

13. Technical reviews of many facets of the Rubber Program can be found in G. S. Whitby, C. C. Davis, and R. F. Dunbrook (eds.), *Synthetic Rubber* (New York and London, 1954). For a brief historical overview of the program, see Dunbrook, "Historical Review," *ibid.,* 32–55. Also see the series of papers by Dunbrook on the contribution of organic chemistry to the rubber program in *India Rubber World* 117 (1947), 203–207, 355–359, 486, 552, 617–619, 745–748. O'Callaghan, *The Government's Rubber Projects,* volume 2, 453–503. Herbert and Bisio, *Synthetic Rubber,* 75–133. Solo, *Synthetic Rubber,* 83–91. Topics covered in detail in later chapters are not referenced in notes for this chapter.

14. Solo, *Synthetic Rubber,* 90.

15. Solo, introduction to *Across the High Technology Threshold,* v–vi.

16. Kurt Meisenburg, German Patent, 757,604, filed 3 March 1937.

17. C. R. Johnson and W. M. Otto, "Monomer Recovery in GR-S Manufacture," *Chemical Engineering Progress* 45 (1949), 409. John W. Livingston and John T. Cox, Jr., "The Manufacture of GR-S," in Whitby, Davis, and Dunbrook (eds.), *Synthetic Rubber*, 198.

18. M. S. Kharasch, et al., "The Laboratory Control of Popcorn Polymer," CR 950, 14 January 1946. The central administration of the rubber program circulated several series of unpublished mimeographed reports including CR (Copolymer Research), CPD (Copolymer Process Development)—which became CD (Copolymer Development) in January 1944—and RM (Raw Materials) reports. Pure research, and much of the applied research, appeared as CR reports. Some applied research on GR-S polymerization was classified as CPD/CD. Sometimes reports went out under more than one code, for example CR 952 (from the NBS) was also CD 583 and RM 356. The reports were usually, but not invariably, circulated about one to three weeks after they were submitted. In citations, the "cover date" (the date of circulation) is generally used and this convention is followed here. Much of the material in the wartime reports was summarized in papers published in 1946–1948, and references will usually be given to these publications. From 1946 onward, the more important CR reports were published more or less simultaneously in the scientific literature, and the scientific papers will be cited in preference to the reports, which are less readily obtained. Most of the original CR, CPD, CD, and RM reports have been collected by the University of Akron archives, and there is a set of CR reports at the University of Illinois archives. The National Archives has only an incomplete set of these reports. They are also available on microfilm as PB 118310 (CR 1–3570), PB 11831052 (CR 3571–3934), PB 11831053 (CR 3935–3962), and PB 126248 (CD 1–3399) from the Library of Congress, Photoduplication Service, Washington, D.C.

19. The next two sections are covered in O'Callaghan, *The Government's Rubber Projects*, volume 2, 539–566. Herbert and Bisio, *Synthetic Rubber*, 135–146. Greer interview; and a private communication from P. S. Greer to Peter Morris, 18 December 1986, hereafter called "Rubber Program Notes."

20. Greer "Rubber Program Notes," 2.

21. Interview of J. Roger Beatty by Peter Morris, 4 November 1986.

22. B. Sheldon Sprague, "An Industrial Innovation That Was Nearly Shelved," *Research Management* (May–June 1986), 26–29. E. J. Powers and G. A. Serad, "History and Development of Polybenzimidazoles," in Raymond B. Seymour and Gerald S. Kirshenbaum (eds.), *High Performance Polymers: Their Origin and Development* (New York, 1986), 355–373.

23. Frank Y. Speight, "Arctic Rubber Research," CR 2112, n.d. Also see Charles C. Price, "How Chemists Create a New Product," *The Chemist* 38 (1961), 131–132.

24. Juan C. Montermoso, "Status of the Army Rubber Research Program," in *Proceedings, Joint Army-Navy-Air Force Conference on Elastomer Research and Development* (Washington, D.C., 1954), 15.

25. Albert W. Meyer, "Problems in Development of Oil Resistant Arctic Rubber," in *ibid.*, 142–145.

26. For the disposal of the plants, see O'Callaghan, *The Government's Rubber Projects*, volume 2, 592–610 and 628–651. Herbert and Bisio, *Synthetic Rubber*, 164–193. Solo, *Synthetic Rubber*, 115–124. Solo, "The Sale of the Synthetic Rubber Plants," *The Journal of Industrial Economics* 2 (1953), 32–43. Charles F. Phillips, Jr., *Competition in the Synthetic Rubber Industry* (Chapel Hill, North Carolina, 1963), 45–62. Stanley E. Boyle, "Government Promotion of Monopoly Power: An Examination of the Sale of the Synthetic Rubber Industry," *Journal of Industrial Economics* 9 (1961), 151–169.

For general accounts of the development of the all-*cis* rubbers, see Frank M.

McMillan, *The Chain Straighteners: Fruitful Innovation: The Discovery of Linear and Stereoregular Synthetic Polymers* (London, 1979), 141–165, and Horne, "The History of Synthetic Rubber," 7–8.

27. Private communication from J. D. D'Ianni, 9 June 1988; information obtained from the Goodyear Information Center.

28. Harry A. McDonald, *Program for Disposal to Private Industry of Government-Owned Rubber-Producing Facilities* (Washington, D.C., 1953), 38.

29. C. E. Jaynes, "Current Air Force Rubber Problems," in *Proceedings, Joint Army-Navy-Air Force Conference on Elastomer Research and Development* (Washington, D.C., 1954), 31.

30. C. S. Marvel, "Final Report to United States Department of Agriculture from the University of Illinois," August 1960, University of Illinois archives (hereafter UI), UI 48/0/1, 1–2.

31. Greer interview, and Greer, "Rubber Program Notes."

32. Table from *Chemical and Engineering News* (30 December 1957), 78, reprinted with corrections in James D. D'Ianni, "Butadiene-Styrene Rubbers," in Maurice Morton (ed.), *Introduction to Rubber Technology* (New York, 1959), 268.

33. Table in *Chemical and Engineering News* (2 June 1975), 33, from information supplied by Rubber Manufacturers' Association.

34. Table in *Chemical and Engineering News* (8 June 1987), 32, from information supplied by Rubber Manufacturers' Association.

35. International Institute of Synthetic Rubber Producers, *International Statistical Review of the Synthetic Rubber Industry* (New York, 1977), Table V: Synthetic Rubber Producers—North America, Estimated Year End 1976 Capacities, 13.

36. Private communication from Herman Schroeder, 6 June 1988. Erik G. M. Tornqvist, "Polyolefin Elastomers—Fifty Years of Progress," in R. B. Seymour and Tai Cheng (eds.), *History of Polyolefins* (Dordrecht, 1986), 152–157. Interview of Charles Price by Peter Morris, 28 May 1987.

37. Kränzlein, *Chemie im Revier*, 113–127. *Buna: Dokumente über ein neues Werk* (Bunawerke Hüls, 1958).

38. Table in *Chemical and Engineering News* (8 June 1987), 32 and 66, the latter from information supplied by the Japanese Ministry of International Trade and Industry.

Chapter 2
Innovation in the Synthetic Rubber Industry

Introduction

Innovation was a major concern of the research program. Solving production bottlenecks and improving the quality of the rubber were important, perhaps more important than innovation during World War II, but this would not have been sufficient reason to extend the research program through 1956. In the postwar period, the focus of the program was on the development of new rubbers that would be competitive with natural rubber. Even the search for new modifiers and monomers, conceived during the war as an "insurance policy" against the failure of the original modifier and styrene as the comonomer, was an example of innovation. Furthermore, most of the current interest in collaboration between government, industry, and academe stems from the premise that such cooperation promotes and accelerates innovation.

The founders of the research program hoped that the whole would be greater the sum of the parts. They assumed that collaboration between different companies, and between companies and universities, would promote innovation. They believed that a free exchange of information and results, unhindered by corporate secrecy and patent laws, would remove barriers to rapid development of these innovations. The government funding was an inducement to participate and an additional means of accelerating technical progress. To use a chemical analogy, they assumed that cooperation would have a synergic effect and the government funding would be a catalyst. Hence, for the government-funded program to be considered a success, it is not sufficient to show simply that innovation occurred. A synergic effect, an acceleration of innovation, has to be demonstrated.

The Corporate Context

The rubber industry was very competitive in the interwar period, in contrast to the chemical industry, and the major rubber companies had their own distinctive character and traditions. U.S. Rubber was oriented toward physical chemistry and physics.[1] Goodrich, on the other hand, was biased in favor of organic chemistry.[2] This bias led to a greater interest in plastics and biologically active compounds, for example, pesticides.[3] Goodrich was also strong in the field of rubber compounding. Its chief compounder, Arthur Juve, became a legend in the industry.[4] Goodyear lay between these extremes.[5] It leaned in the direction of organic chemistry, but also had a strong physics section headed by Samuel Gehman. The different corporate traditions were particularly noteworthy in the field of physical testing of rubber, where each company had developed its own tests and equipment. Of course, even here there were areas of agreement. For example, "Mooney viscosity," developed by U.S. Rubber's chief technologist, was widely used to describe the processibility of a rubber stock.

Given this distinctiveness, each company worked for the most part on its own during the program, taking note of what the other companies (or the universities) had achieved, but developing its own solutions. This was particularly true during the wartime research, when some of the work involved adjusting the basic process to the company's modus operandi and much of it was a matter of trial and error.

Not surprisingly, this distinctiveness was also reflected in the attitude of the big four companies toward the government research program. Despite its prewar research, Goodyear was content to operate fully within the cooperative program, while retaining a small area of private research.[6] Firestone and U.S. Rubber, already collaborating with Standard, also cooperated wholeheartedly with the program, but even U.S. Rubber had a "private" research group led by Frank Mayo.

Goodrich took an even more cautious attitude. The researchers working on the government program were kept away from Goodrich's own research. In the spring of 1943, the group was physically isolated at Kent State University.[7] Goodrich's private ("pioneering") research was carried out by the special projects group, which remained in Akron. There were several conflicts between the two groups, especially over equipment. One researcher who worked at Kent State quipped that they had a better relationship with the other groups in the program than with their own company's R&D department.[8] The two groups were brought into proximity, but not united, in June 1947. Goodrich moved its entire R&D department to Brecksville, between Akron and Cleveland, in 1948. Following the end of World War II, Goodrich lobbied Congress to eliminate the exchange of technical information.[9] The 1948 Rubber Act voided the 1942 patent sharing and information exchange agreements, but the Reconstruction Finance Corporation sidestepped this problem by drawing up new contracts with the various companies and universities, which effectively restored the *status quo ante*.[10]

Phillips Petroleum was not a member of the rubber research program, until its postwar innovations brought it into competition with the other companies. The Reconstruction Finance Corporation then insisted that Phillips's research be incorporated into the cooperative program, with the consequent loss of freedom of action only partly compensated by government funding.[11] Other outsiders were General Tire (a smaller company than the Big Four), and the Polymer Corporation, a Canadian government-owned company, which was more or less expelled from the program at the end of the war since it was a foreign company. It was then obliged by the Canadian munitions minister, Clarence D. Howe, to operate on a "profit or perish" basis.[12] The Polymer Corporation also pursued some interesting research opportunities independently of the companies in the program, including cold rubber and oil-extended rubber.

The Wartime Research

The "mutual recipe" was the cornerstone of the wartime rubber production program. In November 1940, Harvey Firestone, Jr., had suggested that each company produce whatever synthetic rubber it liked, but with the proviso that it use that rubber in its own operations.[13] In March 1942, however, a committee of technologists from the rubber companies, under the aegis of the Rubber Reserve Company, decided to adopt a uniform recipe for the whole industry. All recipes for the production of a GR-S type rubber had three basic components, in addition to the monomers and water: an initiator to start the polymerization (sometimes incorrectly called a catalyst), a modifier to control the polymerization, and an emulsifier to keep the sparingly soluble monomers in emulsion. In the mutual recipe, the initiator was potassium persulfate, the modifier was Lorol mercaptan, and soap flakes were used as the emulsifier.[14]

While the mutual recipe produced a standard rubber and aided the cooperative research, it was not an entirely good idea. Goodrich and Goodyear spent the opening period of the program becoming accustomed to its use, having developed their own formulae. Furthermore, adoption of a uniform recipe, understandable in itself, inhibited the development of new formulae until after the war's end, thereby slowing innovation.

Before the way in which the modifier altered the course of polymerization was fully understood, there were hopes of a major breakthrough with a completely new modifier. But serendipity had done a good job. None of the many compounds tested represented a marked improvement over Lorol. Early results showed the main ingredient in Lorol to be dodecyl mercaptan (DDM), but the impurities present, mainly other long-chain mercaptans, made Lorol a better modifier than pure DDM. Goodrich had developed Di-isopropylxanthogen disulfide ("Dixie") in conjunction with Phillips Petroleum, but it was never popular.

From the beginning of this program, the disappearance of the modifier during the polymerization was a major topic of research. The rubber companies experimented with continuous and stepwise addition of Lorol, and the importance of the amount of stirring used was soon appreciated.[15] Hitherto, vigorous agitation had been used, which hastened the consumption of the modifier. By October 1943, Benjamin Zwicker of Goodrich had produced a constant viscosity polymer by continuous addition and careful control of the stirring rate, and the other companies were not far behind.[16] Shortly afterward, however, the Goodyear group, headed by H. Judson Osterhof and Alvin Borders, discovered that while the viscosity was controlled in this way, the molecular weight distribution was practically the same as before.[17]

Meanwhile, Piet Kolthoff, one of America's best electroanalytical chemists, was improving the determination of low concentrations of mercaptans. Having developed an electrochemical method, he studied the kinetics of the modification process.[18] He concluded that the role of diffusion was an important factor in the reaction. This finding was supported by work carried out by William Reynolds at the University of Cincinnati.

Taken together, their work revealed that tertiary mercaptans diffuse through soap solutions more rapidly than primary mercaptans like DDM. Thus, the controlling factor for DDM was the rate of diffusion, not the rate of reaction. This explained why the rate of disappearance was affected by agitation and the method of addition. Tertiary mercaptans were more nearly reaction rate controlled, and the degree of agitation was relatively unimportant. Furthermore, although the tertiary mercaptan disappeared at a slower rate, less mercaptan was required.

Kolthoff's most significant contribution, however, was the his explanation of Borders's unexpected observation. He showed that while the average molecular weight was related to the amount of modifier added, the molecular weight distribution was controlled by the relative rate of the two steps of the modification process (the transfer constant). The tertiary mercaptans had a better transfer constant and hence produced more middle-range material, and less of the undesirable low and high ends.[19]

This research is a good example of the cooperative nature of the program, with contributions also being made by Robert Frank and Paul V. Smith at the University of Illinois.[20] Ironically, this work hardly affected the commercialization of the tertiary mercaptans, although it may have speeded its acceptance by the industry. Phillips Petroleum had first produced the tertiary mercaptans in 1943 simply because they were easily made from refinery gases.[21] Even after Kolthoff and Reynolds had presented their findings, some processers argued the tertiary mercaptans produced a poorer rubber than DDM.[22] By 1945, using empirical methods, the industry had created (with DDM) as good a rubber as could be made at 50°C.

The problem with the induction period at the beginning of the polymerization had not arisen in the preprogram work, because the soap used had come from a few sources. Indeed, most researchers used the readily available Ivory soap flakes sold by Procter and Gamble. When full-scale production began, however, the soap had to be drawn from a number of sources and obviously the various soaps differed in composition. Variations were found even when the soaps came from one manufacturer.[23]

The plants and the soap companies cooperated to track down the source of the problem, under the active leadership of the Office of the Rubber Director and the Rubber Reserve Company. A considerable volume of experimental work on this problem was also carried out by Marvel's group at the University of Illinois in early 1944.[24] Soaps made from a single fatty acid (soap is usually a mixture of several acids) were used in the polymerization mixture and their effect on the rate of polymerization measured. The saturated acids produced more or less the same amount of polymer in the first twelve hours, but a soap containing 10% of the polyunsaturated linolenic acid only produced half as much. If the polyunsaturated acids in a tallow soap were chemically converted into saturated acids by hydrogenation, a process used in the making of margarine, the rate of polymerization reached optimum. As a result of this research, the Rubber Reserve Company tightened its specifications for the soap.

The situation was complicated by a growing shortage of soap derived from coconut oil and even high-grade tallow. Mildly hydrogenated tallow became the most common source for the soap.[25] This shortage also encouraged research on rosin soaps. In addition to alleviating the pressure on conventional soaps, it was also hoped that the rosin soap residues might improve the finished rubber, especially the tackiness which was poor in GR-S. Rosin soaps had actually been used by compounders before the war to improve the tack of natural rubber.[26] While promising results were obtained, it was found that phenolic impurities in the soap slowed the rate of polymerization. Hercules developed a purified rosin soap, but it did not come into general use until the discovery in 1946 that cumene hydroperoxide (in contrast to potassium persulfate used in the mutual recipe) was insensitive to the presence of rosin soaps.[27]

Once again, it has to be stressed that while cooperation was a feature of the soap research, it was probably inessential to its successful outcome. The tracing of the initial problem to the polyunsaturated acids was not difficult, and the rubber companies were investigating the rosin soaps.

Carbon black is essential for a butadiene-styrene copolymer tire; it increases resistance to wear by the interaction between the styrene rings and the graphite microcrystals in the carbon black. By happy chance, Joe C. Krejci of Phillips Petroleum had developed between 1937 and 1940 a new process that produced the high abrasion furnace-blacks (HAFs), which were particularly suitable for GR-S as they made processing easier and prevented crack growth.[28] The first plant went into operation in December 1943. "By the mid-1950s the replacement of [the older] channel blacks by reinforcing grades of the new oil furnace blacks was almost complete."[29]

The development of furnace-blacks illustrates two general points about the wartime program. Like the tertiary mercaptans, they were developed by Phillips Petroleum as a new outlet for their basic product, petroleum. They were another example of how the foundations of the industry were established before the program began.

Cold Rubber and Oil-Extended Rubber

If a more active recipe were used to polymerize butadiene and styrene, it could be used to make the polymerization at the wartime standard temperature (50°C; 122°F) more rapid, or to carry out the polymerization in the same time (more or less) at a lower temperature. Before a "cold" rubber was possible, it was necessary for the producers to agree that a lower temperature was desirable. Otherwise the higher throughput and the ability to use less pure monomers would favor operation at the original temperature. A report produced in August 1943 by the Committee on Revision of GR-S Formula to Include a Catalyst Activator remarked:

It is the desire of Dr. Williams . . . that the recommendations for an activated recipe be such as to insure the production of improved copolymers. This objective can be attained with activators only if the polymerization is conducted at a lower temperature. This calls for a considerable revision of attitude on the part of some of us.[30]

The advantage of operating at lower temperatures was the decrease in branching and cross-linking, leading to longer linear chains that would have increased strength. Research by W. E. Messer and L. H. Howland (of U.S. Rubber) showed, as early as May 1943, that GR-S produced using the mutual recipe at 30°C significantly improved the tensile strength but increased the polymerization period from twelve hours to sixty-five hours.[31] Clearly, the beneficial low temperatures could only become acceptable if a means of greatly accelerating the rate of reaction could be found.

A prototype recipe had been offered to the Rubber Reserve Company by B. F. Goodrich in December 1941.[32] Instead of the persulfate initiator used by the mutual recipe, it used a mixture of hydrogen peroxide, sodium pyrophosphate, and two transition metal salts (ferric sulphate and cobaltous chloride) in very low concentrations. This early redox catalytic system, with iron in a high oxidation state and cobalt in a low one, was invented by William D. Stewart. He was interested in biochemistry (a characteristic bias of Goodrich research) and derived the concept from a similar process that occurs in muscle.[33] This relatively crude system nevertheless permitted polymerization to take place (to 90% conversion) in twenty hours at 40°C. The sheer novelty of the Goodrich formula militated against its acceptance at the beginning of the rubber program. The need for assured success ruled out any formula unfamiliar to the majority of the participating companies. After all, the Goodrich formula might well have been a failure on the industrial scale.

Messer and Howland replaced the Dixie used by Goodrich with the DDM of the mutual recipe, which further reduced the polymerization period to thirteen hours and 73% conversion at 30°C.[34] Thus, they had succeeded in reproducing the standard conditions of time and conversion at a significantly lower temperature. Goodrich experimented with the addition of a further reducing agent (DDM acted as reducing agent in the mutual and Messer-Howland formulae). When small amounts of hydroquinone were used, the time required at 30°C was reduced to ten hours.[35] Goodyear was not so active in this line of research, but developed a perborate and cobalt formula which was used to produce the company's "private" butadiene-acrylonitrile rubber at 40°C.[36]

A considerable body of research was also carried out during the war on recipes activated by potassium ferricyanide, a catalyst first used by Howard Starkweather, Sr., at Du Pont to produce neoprene.[37] The California Development Program, based at the Goodyear plant at Torrance, California, near Los Angeles, was the most active center of development.[38] The original formula used caustic soda, and there was considerable concern about the effect of this highly alkaline solution on the "beer" glass—so called because it was a glass similar to the beer bottles used in laboratory experiments—used to line the reactors.[39] In early 1944, however, the Torrance group introduced a caustic soda-free recipe that replaced pure potassium ferricyanide with a cheap commercial mixture of sodium and potassium ferricyanide (Redsol).[40] In March 1944, the Rubber Research Board considered using a ferricyanide-activated formula to reduce the polymerization time at 50°C, and hence produce a sorely needed increase in GR-S production. Instead, it decided to order extra reactors and to concentrate on improving the formula so it could be used at lower temperatures.[41] In 1944, low-temperature polymerization meant polymerization at 40°C, not 5°C! The main problem was the insufficient supply of potassium ferricyanide or even the more readily available Redsol.

This reluctance to introduce the new process was increased by doubts about the reality of the improvements produced. These doubts were eroded by the program's increasing understanding of the structure of synthetic rubber, cross-linking, and the importance of a high molecular weight. Many of the doubts entertained by the Rubber Research Board in January 1944 were removed a few months later when the results of U.S. Rubber's tests on GR-S polymerized (without a catalyst) at 30°C were completed. The properties of this rubber were described as outstanding.[42] At the last meeting of the Rubber Research Board in July 1944, John W. Livingston of the Rubber Reserve Company agreed to raise with the Operators' Committee the issue of converting some reactors to operation at 40°C.[43]

Nothing apparently came of this initiative, but the improving situation for the low-temperature recipes was strongly reinforced by the shattering news that the Germans were ahead of the Americans in this field.[44] The initial reports were brought back in the summer of 1945 by Livingston and further information was gathered that fall by Carl Marvel. The Germans had succeeded in carrying out the polymerization in twelve hours at 10°C or in less than an hour at 40°C. The key ingredients in the recipe which made this possible were the use of sugars (for example, sorbose) as an additional reducing agent, an organic peroxide (benzoyl peroxide) as the oxidant instead of hydrogen peroxide, and the existence of reducing agents in the synthetic emulsifier the Germans used in place of soap.

Armed with this new knowledge, the American researchers sought an optimum recipe. By the spring of 1946, Piet Kolthoff and Edward Meehan had developed a rosin soap recipe which used glucose and cumene hydroperoxide.[45] Hercules was studying cumene hydroperoxide, formed by the action of air on cumene (isopropylbenzene)—research that led to the Hercules-Distillers process of phenol manufacture from cumene in the early 1950s—and suggested that it could be used as an initiator.[46] The ferric sulphate of the earlier American formulae was replaced by the ferrous sulphate used by the Germans. Employing a formula similar to this, Charles Fryling of Phillips Petroleum was able to show that there was a marked improvement in properties in lowering the temperature from 50°C to 15°C.[47]

The scene now shifted to the industrial pilot plants. Phillips Petroleum was particularly well placed to develop cold rubber. The company produced (in addition to butadiene) tertiary mercaptans and carbon blacks which could be tailored for use with cold rubber. Phillips hired Charles Fryling in 1945 and Professor William Reynolds of the University of Cincinnati in 1946. Both men were familiar with cold rubber: Reynolds had collaborated with Kolthoff, and Fryling had worked with Stewart at Goodrich. Tires made by Phillips were supplied to the Office of Rubber Reserve (ORR) for testing by the government tire fleet in October 1947.[48]

The results were so impressive that the ORR authorized the Copolymer Corporation (a consortium of small rubber companies) to convert half its Baton Rouge plant to the production of cold rubber. The new plant started up in February 1948. The research at Phillips and Copolymer led to the adoption of the "Custom" recipe, which was broadly similiar to Kolthoff and Meehan's formula, except that the DDM was replaced by Phillips's sulfole tertiary mercaptan. The polymerization was carried out at 5°C for practical reasons. The recipes for operation at lower temperatures, even down to 0°F (−18°C), already existed, but such low temperatures required special equipment and the use of methanol as an anti-freeze.

During the summer of 1948, the RFC was under unprecedented pressure from the rubber companies and the newspapers to rapidly expand the cold rubber program. The RFC insisted, however, on awaiting the final results of using the new rubber in selected factories. In October 1948, the RFC declared that cold rubber was "the best tire tread known," giving "about 30 percent larger mileage in tire treads than previous synthetic rubber."[49] The corporation therefore said that it would convert eight plants to cold rubber, increasing the capacity from 21,000 to 183,000 long tons. Not surprisingly, the Copolymer Corporation plant was the first to be completely converted to cold rubber. This event was celebrated by an "open day" on 24 May 1949. By the end of that year, half of all GR-S capacity had been turned over to cold rubber production.[50] In 1954, two-thirds of all GR-S produced in the United States was cold rubber.[51]

The outbreak of the Korean War in the summer of 1950 shifted the focus of the program. The cost of synthetic rubber was no longer a problem, but there was a marked shortfall in supply as capacity had been cut back since the end of World War II. While mothballed plants could be (and were) brought back into use, the rubber they produced was excessively expensive. But a solution to these problems was at hand.

A rubber can be made to go further by the addition of cheap, inert "extenders." All rubbers, natural and synthetic alike, are extended rubbers, since they contain inorganic fillers such as carbon black and zinc oxide. Far from detracting from the rubber's properties, they have a strong reinforcing effect and make the rubber cheaper. Natural and artificial pitches are often used to extend natural rubber. A rubber can also be made more processible by adding "plasticizers." Buna S (or GR-S) can be plasticized by direct scission of the chains by the action of oxygen ("thermal degradation") or by modifiers. A further possibility in the case of synthetic rubbers is the addition of "lubricators." These are no longer regarded as being distinct from plasticizers, but during the early development of synthetic rubber the lubrication of a "dry" crumbly rubber was often considered. Indeed, Buna S was a direct result of an attempt to lubricate "dry" emulsion-polymerized polybutadiene with linseed oil.[52]

Until 1943, the Germans produced a very tough polymer which was then heated at 130°C to break down the long chains; this permitted easier processing but badly weakened the rubber.[53] The American industry used modifiers to produce the same results, but without the excessive formation of the gel that gave rise to the poor final properties. The German industry switched to modified rubber in 1943, but a high molecular weight was still preferred by I.G. Farben. Within a few months of the introduction of this Buna S3, the Austrian rubber company Semperit began to extend it with mineral oil because of the desperate shortage of rubber. While the properties of Buna S3 suffered as a result of this treatment, it was also much easier to process.[54] The Germans also used a low molecular weight polybutadiene (Buna 85) as a plasticizer.

The possibility of using plasticizers or extenders was by no means ignored in America. Goodrich published the results of a survey of 650 possible softeners in 1944,[55] and a similar survey of 110 extenders the following year.[56] In both cases, the research group concluded that no single softener or extender could be recommended, because the requirements of the finished product determined which one should be used. Nevertheless, it was reported that GR-S could be extended up to 20% without serious loss of the physical properties. It is interesting to note that mineral oils were the only group common to both softeners and extenders. This line of research was not taken up, largely because the postwar focus of the research program was improving the quality of

GR-S to meet the challenge of natural rubber, rather than increasing the quantity.

Emmet Pfau, a young chemist at Goodrich (not connected with the group mentioned above), was developing the idea of replacing the "soupy" low molecular weight polymer in GR-S with oil, which would be a far cheaper plasticizer.[57] His original intention was to displace the modifier altogether. It was not possible to bring this concept to complete success with "hot" GR-S because driving the polymerization to a high enough weight led to a polymer with excessive branching and cross-linking. The molecular weight distribution was also too broad. These difficulties were removed with cold rubber. It was now possible to produce a linear, gel-free, high molecular weight rubber. Furthermore, the use of tertiary mercaptans restricted the spread of molecular weights to a narrow band. Indeed, the absence of the "soupy" low ends made the addition of a plasticizer such as oil advantageous, leading to easier processing. It also increased crack resistance and reduced heat build-up, two major problems with GR-S.

Pfau and his coworker Harold Brown were not able to interest the Goodrich management in the process. Pfau left Goodrich in 1949 and joined General Tire. General Tire was keen to develop the process with a company working on cold rubber and, if possible, to develop it outside the United States to avoid legal complications with the RFC. The Polymer Corporation met both these requirements, and most of the large-scale development was carried out at Sarnia, Ontario, in 1950 and 1951. The major problem was the selection of an oil that had the best effect on the rubber and also stayed in place. The quality of the oil was discovered to be a key factor. By November 1950, William O'Neil, General's president, was sufficiently confident of the outcome of this research that he offered it to the RFC to bridge the rubber shortage created by the Korean War. He was not particularly forthcoming about the details because he was concerned about the lack of a patent to protect his company's process. Under the circumstances, the RFC's technical advisors did not feel able to recommend that RFC buy this "pig in a poke."[58]

Goodyear announced five months later that a team headed by Osterhof had also perfected oil-extended rubber. There were no complications in this case, as Goodyear willingly agreed to treat this development as part of the work shared in common. The commercial production of oil-extended rubber by Goodyear was publicized by the RFC in February of 1951.[59] General immediately cried foul, claiming that Goodyear had stolen its idea. How, in fact, did Goodyear come by oil-extended rubber? It is possible that Goodyear heard about General's work through a third party. Conversely, D'Ianni has claimed that Goodyear had been looking at materials that could be used to extend rubber because of the Korean War.[60] From a historical perspective, the idea of using oil to extend rubber was scarcely a conceptual breakthrough, but the whole story will probably never be known.

A common feature of the General and Goodyear work was the addition of the oil to the latex (rather than the solid rubber), a process known as oil masterbatching. The idea of adding the other ingredients such as carbon black to the latex rather than the solid rubber had been considered for natural rubber.[61] However, the distance between the plantations and the pigment and rubber factories militated against its practical application. I.G. Farben applied the principle to synthetic rubber in the late 1920s, obtaining a U.S. patent in 1935.[62] Latex masterbatching of carbon black is useful in the case of synthetic rubber because the various manufacturing plants involved are often relatively close, and because synthetic rubber places an extra strain on the mixing machinery used for solid rubber. Power is also saved, and the dispersion of the carbon black in the rubber is improved.[63]

Although all the rubber companies in the program were interested in carbon black masterbatching, much of the early work was done by General Tire and U.S. Rubber.[64] Goodyear started using the process at its Houston, Texas, plant in the early postwar period. In 1950, before oil-masterbatched rubber was introduced, carbon black–masterbatched rubber accounted for 18% of all GR-S.[65]

The masterbatching process was particularly suitable for the addition of oil extenders, and this was one factor that encouraged General Tire to develop oil-extended rubber. Oil and carbon black–masterbatched rubber and oil-masterbatched rubber came on the market in 1951. By 1953, oil-masterbatched rubber accounted for 18.5% of all GR-S, and oil black–masterbatched rubber for an additional 2.5%.[66] The consumption of oil masterbatch increased rapidly, especially in tire treads. It now accounts for half of all butadiene-styrene rubber (SBR) in the United States: three tons of oil masterbatch is produced for every two tons of non-masterbatch SBR.[67]

Since activated recipes and the value of low-temperature polymerization were known in 1943, why was the development of cold rubber delayed until late 1945? The suggestion made by the I.G. Farben chemists, that sugar be added to the recipe, was useful but hardly essential, as sugar-free recipes were introduced in the early 1950s. The other elements of the German recipe, hydrogen peroxide and iron compounds, were already used in the Goodrich formulae. Furthermore, William Reynolds and Piet Kolthoff developed excellent activated recipes based on different principles (they used potassium ferricyanide and diazothioethers) before the end of the war.[68]

The main reason for the delay was the inflexible nature of the wartime production program. The mutual recipe had been chosen to make it easier for the companies to cooperate and to simplify the processing stages. However, this made it more difficult to test or introduce new recipes on a large scale. It was clearly important not to disrupt the flow of sorely needed GR-S from the production plants. Such weighty reasons provided the perfect alibis for bureaucratic inertia. It is therefore not surprising that Goodrich's and Goodyear's "private" nitrile (Buna N) rubbers became the vanguard of the technology because they did not have to conform to the mutual recipe.

Oil-extended rubber can only be made from cold rubber and hence could not have been introduced before it. Once a gel-free rubber of high molecular weight could be produced, however, the concept of extending it with cheap materials (such as mineral oil) could not be far behind, if only to cure its "dryness." This is illustrated by the use of castor oil by the Polymer Corporation and Jersey Standard to cure the buckling of butyl rubber at low temperatures in 1948, a full two years before General succeeded with oil-extended GR-S.[69]

Why did Goodrich fail to develop the cold rubber process or oil-extended rubber? On the basis of his interviews with Goodrich's research staff, Robert Solo has argued that the company had little interest in GR-S and no interest in developments that could not be exploited for the company's benefit.[70]

While it is true that Goodrich was one of the less accommodating companies in the program, it was probably doing as much as any company would do in the absence of a compelling need to compete. The formulae being used in 1946 were significantly different from Goodrich's original recipe, so it is perhaps not too surprising that the company lost its prewar lead. As soon as the process was taken up by the RFC in 1948, Goodrich began to develop a continuous cold process in collaboration with the Government Laboratory in Akron: by 1950, it was in full-scale production.[71]

Goodrich's rejection of oil-extended rubber is also easy to understand. The extension of rubber with oil was not possible until cold rubber was introduced, and thus after Pfau left Goodrich. Furthermore, there was no compelling need to develop the cheaper oil-extended rubber in the absence of price competition. It should also be noted that Pfau's original idea, that oil could be used to replace the modifier altogether, was incorrect. Oil-extended cold rubber still required a modifier during polymerization.

Intermezzo: Infra-Red Spectrophotometry

The "infra-red" is that part of the electromagnetic (or solar) spectrum which lies in the region beyond the red band of the visible spectrum, and similarly, the "ultra-violet" lies beyond the violet end. Just as the elements exhibit dark absorption lines in the solar spectrum (the Fraunhofer lines), organic compounds exhibit broad absorption lines—usually called bands—in the infra-red region. Obviously the human eye cannot detect these bands, but they can be recorded by machines called spectrophotometers. Each band is associated with a specific type of molecular vibration in the organic compound, such as that of a carbon-carbon bond or a carbon-hydrogen bond.

The use of infra-red spectroscopy as a means of identifying organic compounds first came into use in the 1930s. The first commercial instruments were produced during World War II, in conjunction with the rubber program by National Technical Laboratories (now Beckman Instruments), and independently by Perkin-Elmer.[72] Robert Brattain at the Shell Development Company at Emeryville, California, collaborated with Arnold O. Beckman, E. D. Haller, and Kenyon L. George of National Technical Laboratories in the development of its first series of commercial instruments, from IR–102 through IR–107.[73] Infra-red (and ultra-violet) spectrophotometry was developed during World War II by Shell Development Company, Humble Oil Company, and Phillips Petroleum as a rapid method of analyzing gaseous feedstock streams during the manufacture of butadiene. A summer school in spectrophometric analytical methods, led by Brattain, was sponsored by the Rubber Reserve Company and the Petroleum Administration for War in August 1943.

The use of infra-red spectrophotometry for the analysis of solid synthetic rubbers was slower to develop. Goodyear first entered this field when they hired John Field of the University of Michigan in 1936 on the condition that he bring his self-constructed instrument with him.[74] A research group at American Cyanamid, headed by R. Bowling Barnes, published a discussion of the infra-red spectra of the monomers and polymers involved in making synthetic rubber in 1943.[75] Unfortunately, they did not examine the important region of the spectrum between 700 and 1000 cm^{-1}.

In 1943, the ORD circulated a report written by John White and Paul Flory on work clearly carried out while Flory was still working for Jersey Standard (i.e., about a year earlier).[76] White and Flory used infra-red spectrophotometry to confirm the chemical structure of polyisobutylene, and the copolymers of butadiene with styrene and acrylonitrile. They were able to show, for example, that the cyanide group of acrylonitrile remains intact in the copolymer. They also realized that the spectra could be used to measure the proportion of external double bonds in the butadiene polymers (the 1,2 content). They correctly concluded that sodium-polymerized polybutadiene had a considerable 1,2 content, but they rather underestimated the 1,2 content of the butadiene-styrene copolymer. In any case, the absorption bands were too strong to permit quantitative estimates.

In 1946, Samuel Gehman, John Field, and D. Woodford at Goodyear used the relative intensity of the bands at 996 cm^{-1} and 967 cm^{-1} to measure the 1,2 content of GR-S, not realizing the latter band was only produced by the *trans* internal double bond.[77] Finally in 1949, on the basis of earlier work by his colleagues Albert Meyer and Edwin Hart,[78] Robert Hampton of U.S. Rubber used an elegant set of simultaneous equations to estimate the 1,2 content and the *cis/trans* content of synthetic rubbers.[79] Using pure hydrocarbon samples from the National Bureau of Standards, Hart, Meyer, and Hampton had allocated the 724 cm^{-1} band to the *cis* internal double bond, a band hitherto ignored by researchers. They used the 910 cm^{-1} band to measure the 1,2 content, instead of the 996 cm^{-1} band used by the Jersey Standard and Goodyear groups. The proportion of *trans* internal double bonds was measured using the 967 cm^{-1} band misidentified by the other groups as common to both internal double bonds. Hampton was now in a position to calculate the *trans* content of any synthetic rubber, and to discover the effect of, say, the polymerization temperature on a polymer's structure. For example, he was able to show that the *trans* content of polybutadiene fell with increasing polymerization temperature, whereas the 1,2 content remained constant. This convenient method of measuring the *cis* content of synthetic rubber greatly assisted the development of an all-*cis* synthetic "natural rubber," although the diffusion of Hampton's technique was apparently slow.[80]

Synthetic "Natural Rubber"

The development of synthetic "natural rubber" was connected with the greatest failure of the government research program: the all-synthetic rubber large truck tire (and the related airplane tire). Truck tires usually operate at high temperature under heavy loads. GR-S tires suffered high heat build-up while running, and this problem was accentuated by the synthetic rubber's poor strength at high temperatures. A synthetic rubber with a low degree of internal friction (hysteresis), which produces heat, was required. Goodrich's John Collyer had described the truck tire as a major problem in testimony to a Senate committee at the beginning of 1942, and the firm hoped to solve it before the end of that year.[81] The problem persisted, however, and while cold GR-S was very successful in passenger-car tires, the official history noted in 1955 that the all-synthetic heavy duty tire was the program's major unattained goal.[82]

In common with several other firms, including Goodyear and Du Pont, Goodrich had high hopes in 1950 for polyurethane rubber (Estane). It was an excellent rubber in the laboratory with a promising hysteresis, but road tests soon revealed that tires made from Estane melted on skidding. As Herman Schroeder has remarked of the similar Du Pont rubber (Adiprene), it "was fine as long as you didn't have to stop fast somewhere along the way!"[83] Goodrich's private American Rubber Team investigated new copolymers and terpolymers. Initial results suggested that acrylates and methacrylates could reduce heat build-up in tires.[84] Nirub C, a terpolymer of butadiene, isoprene, and methyl methacrylate, was described as having "outstanding low hysteresis properties."[85] Unfortunately, the laboratory chemists had not interpreted the test results correctly and further testing showed that these new rubbers did not offer any significant improvement over GR-S. Nirub C, for example, was found to be brittle and cracked when made into tires.[86]

It soon became clear that only synthetics that had an all-*cis* structure were likely to equal natural rubber for heavy duty tires, and perhaps only all-*cis* polyisoprene. The I.G. Farben chemists had stressed the value of an all-*cis* configuration to American investigators in the fall of 1945.[87] The potential of this structure was also recognized by Bell Telephone Laboratories.[88] Goodrich apparently became aware of the importance of the *cis* configuration in May 1950.[89] This ruled out the copolymers, such as cold rubber. There had been interest in polybutadiene and polyisoprene since the beginning of the program—a reflection of the general distrust of styrene—but they required different polymerization systems. Emulsion polymerization of butadiene produced a weak, "dry," highly cross-linked polymer.

Marvel and his students, notably William J. Bailey, improved the sodium-catalyzed bulk polymerization of butadiene—discovered by F. E. Matthews in 1910—by adding dilutants, for example, toluene, and by replacing sodium pellets with sodium "sand." This "sand" was made by high-speed agitation of molten sodium in toluene. Although not all-*cis,* the sodium-polymerized polybutadiene and the butadiene-styrene copolymer were superior in heat build-up to GR-S, and even natural rubber. Mixed with natural rubber, these polymers could be made into a good tire. In tests of truck tires, made with 68% synthetic and 32% natural rubber, the sodium-polymerized copolymer tire ran about twice as far as the GR-S tire. This was not, however, a commercial success, largely because of its poor low-temperature properties.[90]

The American Rubber Team reexamined sodium polymerization in 1950–1951 and, as the original polymers had excessively high molecular weights, looked for a "modifier" that would reduce the chain length. Ethylene glycol monoethyl ether was found to be an "outstanding" modifier—research chemists apparently have a fondness for this particular adjective—but the sodium rubbers were eventually abandoned. The team's progress report for August 1951 commented:

Because 13 months of work on the sodium catalyzed butadiene polymers has failed to yield anything really new or promising for the carcass program, and because it has not been possible to find a field which does not fall under the government program, this part of the rubber team program is being discontinued.[91]

Another route to the homopolymers, the "Alfin" catalyst, was developed by Avery A. Morton of the Massachusetts Institute of Technology.[92] The catalyst was an insoluble aggregate of sodium chloride (salt), sodium isopropoxide (made by the action of sodium on isopropyl alcohol), and allyl sodium. The name "Alfin" is a contraction of *al*cohol and ole*fin* (the propylene used to make allyl sodium). The polymerization took place in a solvent, usually pentane. One major problem was gel formation by reaction between the catalyst and the polymer, so the catalyst concentration was kept low. It was thought possible at the time that the Alfin polymers might be significantly higher in *cis* content than the random polymer, but it was eventually discovered to be higher in *trans* content. The products had high molecular weights because it was impossible to "modify" the reaction, and this made processing difficult. This high molecular weight did produce some improvement in abrasion resistance over GR-S. Nevertheless, the erratic activity of the Alfin catalysts and the problems with gel formation more than counterbalanced this small potential improvement.[93]

It is remarkable that there was so little interest in the work of Karl Ziegler, which was accessible in the open literature. He had first made polybutadiene with butyl lithium in 1934 and had continued to work on this polymerization during the 1930s.[94] When he reacted a solution of butadiene in ether with metallic lithium (sodium) in the presence of amines, the addition was halted at the first step. On working up the solution, *cis*–2-butene was obtained in a quantitative yield.[95] In 1940, he polymerized isoprene with lithium.[96] It is particularly surprising that Carl Marvel did not pursue this line of research, as he had been interested in Ziegler's work as early as the mid-1920s and was one of few American chemists familiar with butyl lithium. While Marvel did carry out some experiments with lithium polymerization, he believed that the lithium polymers offered no advantage over the sodium analogues. Other researchers were even less successful. Thomas Midgley remarked in 1937 that the polymerization of isoprene with lithium produced "zero results." Goodrich later found lithium to be an "ineffective" catalyst for the polymerization of butadiene.[97] These negative results probably stemmed from the use of impure materials.

Before 1952, the most successful American research with lithium polymerization was the polymerization of butadiene with lithium hydride by Morris Kharasch in 1944.[98] He even found a modifier for this polymerization, thereby avoiding the high molecular weight polymers produced by the Alfin process. His polymer was almostly certainly a high-*cis* polymer, but the significance of producing such rubbers was not yet fully appreciated in 1944. Kharasch also lacked a simple means to identify a high *cis* content. His former student Cheves Walling noted this near-miss in his 1959 account of Kharasch's work and commented:

Undoubtedly, more complete evaluation of these differences [between Kharasch's polymer and other polybutadienes] was handicapped by the absence of the rapid methods for the determination of polymer structure (such as infrared spectroscopy) which are now available.[99]

In 1952, Frederick Stavely, another former Kharasch student, who was now research director at Firestone, initiated a major effort to produce an all-*cis* rubber. Stavely decided to concentrate on polymerization with alkali metals, such as lithium and sodium, because it appeared to offer a greater variety of structures. As he noted in his Goodyear medal address:

Emulsion polymers had predominantly *trans*-1,4 structure whereas sodium catalyzed polymers of isoprene contained about equal amounts of structures obtained by 1,4 addition (32%), and 1,2 addition (32%) and about 24% by 3,4 addition as reported by Dr. J. D. D'Ianni of . . . Goodyear. . . . Polyisoprenes made by the Alfin catalyst had much higher *trans*-1,4 structure.

The work by Marvel and Morton had indicated the potential of solution polymerization, but Stavely decided to pursue the bulk polymerization of isoprene using metallic lithium. He chose isoprene, and not butadiene, in case the extra methyl group made a crucial contribution to the superior properties of natural rubber. Lynn Wakefield was able to produce a polyisoprene with an increased *cis* content in early 1953. By September, a pilot plant was in operation.

The second breakthrough occurred when the pilot plant was allowed to run longer than usual, which produced a coral-like rubber with a high (94%) *cis* content. The difficulties involved in digging a solid block of rubber out of the reactor impelled the introduction of solution polymerization. Explosions were common during this stage of development. Further research led to the replacement of lithium metal by butyl lithium. Firestone thus came to the process implicit in Ziegler's papers by a roundabout route.[100]

In the same year that Firestone was building on Karl Ziegler's work, he created a new catalyst system (a combination of aluminum alkyls and titanium chloride) which he used to polymerize ethylene. This polyethylene differed from any previous commercial polyethylene in being completely linear, without branching. Despite his earlier work on synthetic rubber, Ziegler concentrated on the polymerization of ethylene and did not attempt to develop the polymerization of butadiene or isoprene. Furthermore, the West German synthetic rubber industry was just recovering from the postwar ban on synthetic rubber production.[101]

Partly as a result of its success with polyvinyl chloride (Geon), Goodrich was interested in the development of nonelastic thermoplastics. For example, the Technical Service Research Department studied the high-pressure polymerization of ethylene in the early 1950s, producing several copolymers on the experimental scale. In the summer of 1953, Waldo Semon and Carlin Gibbs, head of the Polymerization Research Department, visited Ziegler's laboratory during a European trip. A few months later, Goodrich formed a joint company with Gulf Petroleum to commercialize petroleum-based plastics. A Goodrich-Gulf representative learned of Ziegler's breakthrough in May 1954 through Gulf's connections with Ruhrchemie, one of the sponsors of Ziegler's research. Goodrich-Gulf and Ziegler discussed the possibility of a license to make linear polyethylene, and eventually the company obtained a nonexclusive license to use the new catalyst.

The Goodrich-Gulf research group, headed by Carlin Gibbs, had trouble with the polymerization of ethylene, until the need for very pure ethylene and the complete exclusion of air was established. Gibbs decided to use the catalyst in an attempt to make a copolymer of ethylene and 10% isoprene, which would produce a cross-linkable thermoplastic. The experiment was delegated to Samuel Horne, who had joined Goodrich in 1950. The mushy product was then sent to Goodrich's infra-red specialist, James Shipman, for analysis. He was the first to realize that Horne had made all-*cis* polyisoprene or synthetic "natural rubber"; he thought Horne was playing a joke on him by handing in a sample of natural rubber.

Gibbs was now faced with a legal dilemma. Ziegler had inserted a stiff penalty clause in the license contract, which prevented unauthorized disclosure of his catalyst to third parties. On the other hand, any research by Goodrich on copolymers containing more than 50% of any diene had to be published in a rubber research program report. This problem was initially circumvented by bubbling ethylene through the reactor used to polymerize isoprene, to allow Goodrich-Gulf to claim that it was working on copolymers of ethylene and isoprene. The polyisoprene was then dissolved in benzene to remove any polyethylene formed. A few days later, Goodrich's legal department decided that Goodrich-Gulf was outside the scope of Goodrich's contract with the Reconstruction Finance Corporation (RFC), and this pretense was dropped. Events now moved quickly. Within weeks, the polymerization of isoprene was carried out in a 1,000-gallon reactor. Semon informed Ziegler of Goodrich-Gulf's success, and the new rubber was announced to the public on 2 December 1954.[102]

Alarmed by this news, Goodyear approached Firestone and learned of Stavely's new process. The two companies collaborated for a few months on the industrial development of the Firestone process. Unhappy about being behind the other two companies—and with the presentation of papers about the new rubbers scheduled for the American Chemical Society Rubber Division meeting in Philadelphia in November 1955—the Goodyear chemists decided to do it on their own. Aware that Ziegler had developed a new catalyst, but knowing nothing of the details of the Goodrich process, they tried a Ziegler catalyst on isoprene. Goodyear thereby succeeded in making synthetic "natural rubber" too. The company stole a march on its rivals by giving the details to the technical press before the Philadelphia meeting.[103]

In the meantime, Shell Chemical Company and Phillips had also, independently, succeeded in making high-*cis* rubber. A field that was empty two years earlier was becoming crowded by the end of 1955. Eventually, Firestone and Phillips concentrated on polybutadiene, while Goodyear and Goodrich continued with the development of polyisoprene. The rubbers produced by Firestone and Shell were high in *cis* content (up to 95%) but were not all-*cis,* and this turned out to be a serious problem, with the result that both companies left the field. But, at long last, the truck tire problem was solved; even the Firestone polyisoprene was practically as good as natural rubber in tire tests.

In normal times, however, natural rubber has continued to dominate this sector, as isoprene is relatively expensive, even when produced by a Goodyear process based on earlier work by Ziegler's group. Ziegler could now take the credit, not only for the polymerization methods used by all three rubber companies, but also for the synthesis of the monomer as well. Of course, synthetic "natural rubber" would be of immense value (and even cheap) if natural rubber were cut off or severely limited for any reason. The industrialized countries now have no reason to fear an OPEC-style rubber cartel, as they did in the 1920s.

Under their government contracts, the managements of Firestone and Goodrich were obliged to inform the RFC of their work and to license other rubber companies without charge, even if the government was not the source of the funds used. This appears to be the reason for the lukewarm attitude of Firestone's top management toward Stavely's results, rather than its inability to understand the importance of the results. Even when it knew that Firestone had been bettered by Goodrich, the management delayed announcing the new process until August 1955, after the government had sold the plants. Nevertheless, Firestone agreed to honor its contract by publishing its discovery and thereby placing it in the public domain. If the company had done otherwise, it would have later been in the embarrassing position of suing another company (General) for doing the same thing with oil-extended rubber. The government's lawyers disputed Goodrich's contention that Goodrich-Gulf was not obliged to disclose its research to the RFC. The RFC made preparations to bring the case to court, but the matter was dropped after the Federal Facilities Corporation (legal successor to the Office of Synthetic Rubber) was dissolved in 1961.[104]

Why did it take so long to realize that the lithium-catalyzed polymer could be an all-*cis* rubber? It was not easy to characterize an all-*cis* rubber until U.S. Rubber—who ironically took no part in the development of such rubbers—correctly identified the infra-red bands of the *cis* and *trans* internal double bonds in 1949. However, X-ray diffraction had been used for many years to distinguish natural rubber from most synthetic rubbers.[105] Only natural rubber gives a sharp pattern when stretched. An all-*cis* polymer would give a similar pattern. Furthermore, such a rubber might have been spotted because of its excellent physical properties, as appears to have been the case at Firestone. All potentially useful rubbers were put through a battery of tests by the rubber companies or the Government Laboratories at Akron.

Marvel's limited success with the sodium-catalyzed rubber may provide a partial explanation. If the sodium-based rubber (or the similar Alfin rubber) had been clearly superior to GR-S, this breakthrough would have stimulated more research on metal-catalyzed polymerization as a class. As it was, Marvel and Morton were occupied with the improvement of their sodium-based processes (and in Marvel's case, many other things beside). It is likely they thought that the lithium-catalyzed polymer would be similar to the sodium-polymerized rubbers. Furthermore, the idea that a catalyst as simple as lithium or even butyl lithium would produce a rubber with a highly ordered structure seemed improbable. In this respect, the polymers produced with the Ziegler catalysts were less surprising than the lithium-catalyzed polymers.

Faced with the apparently daunting task of finding a suitable catalyst, most workers in this field had used complicated routes to all-*cis* polymers. For example, Goodrich tried three different solutions in 1950 and 1951. Butadiene was polymerized in the presence of complexing agents, such as silver nitrate, which might influence the polymerization process. These agents altered the infra-red spectrum, but the spectroscopists were unable to tell if the *cis* content had changed.[106] Condensation polymers were prepared from organic acids with a *cis* configuration, but they were not very successful. The polymers prepared from citraconic anhydride produced poor inelastic fibers. It was impossible to prepare esters of mucaconic acid (1,4 butadiene-dicarboxylic acid) from the free acid.[107] The most imaginative solution was the proposed formation of polyisoprene from a copolymer of ethylene and isopropenyl acetate by the elimination of acetic acid:

$$O = C \overset{\diagup CH_3}{\diagdown O}$$

$$\cdots - CH_2 - CH_2 - \underset{\underset{CH_3}{|}}{C} - CH_2 - \cdots$$

$$\downarrow$$

$$HO - C \overset{\diagup CH_3}{\underset{O}{\diagdown\!\!\diagdown}}$$

$$+$$

$$\cdots - CH_2 - CH = \underset{\underset{CH_3}{|}}{C} - CH_2 - \cdots$$

Unfortunately, several attempts to find a suitable synthesis of isopropenyl acetate failed.[108] In any event, it was uncertain if the desired 1:1 copolymer of ethylene and isopropenyl acetate could be prepared. For the rubber companies, especially Goodrich, these intricate routes were blind alleys and absorbed a considerable amount of research effort and funds.

Research was also directed toward a better understanding of the properties of stereoregular polymers, rather than toward the innovation of a new synthetic rubber. William Bailey, who was on the faculty at the University of Maryland after 1951, synthesized several all-*cis* polymers in the early 1950s with the help of a grant from the Office of Naval Research.[109] While he succeeded in synthesizing an all-*cis* polymer with a structure closer to polyisoprene, his routes were not intended to be commercially viable.

Another reason for the delay was a lack of interest by the rubber industry in the various loose ends thrown up by their own research and the university-based research. This flotsam, unpromising at first sight, can nevertheless be a source of serendipitous breakthroughs. Stavely appears to have been the first to pursue this approach in the field of all-*cis* polymers. In the absence of competition, and with an obligation to share information, there was no compelling reason for the rubber companies to follow these leads. The chief advantage of this unorthodox approach is the strong patent position it gives a successful company. This incentive was wiped out by the patent-sharing agreements with the Reconstruction Finance Corporation. Firestone only began to chase up the various clues pointing to lithium polymerization when the end of the program was in sight. Ironically, Stavely's group succeeded too quickly; the patent agreements were still in existence when they made their breakthrough.

This argument also applies to the Ziegler catalysts. Ziegler, who sought close control over the use of his patents, would have been cautious about giving all the rubber companies a license for his catalyst. Yet this would have been the only possible outcome under the exchange agreement. It will be recalled that Ziegler discussed the licensing issue with Goodrich-Gulf, which was outside the program, not Goodrich. Furthermore, the patent agreement may have inhibited the program's members from seeking further information from Ziegler after he announced his initial breakthrough in 1950.[110]

The announcement of Horne's discovery was followed by a revival of competition: Firestone and Phillips accelerated their research, and Shell and Goodyear felt compelled to reply. There had not been such frenzied activity on the research front since 1942.

Butyl Rubber

Several of the technical developments that took place after the program ended were connected with butyl rubber. The history of butyl rubber goes back to the early 1930s. I.G. Farben and Jersey Standard were working together on the development of polyisobutylene, first prepared in I.G. Farben's Oppau laboratories in 1931, and which was used by Standard to increase the viscosity index of motor oils. In 1937, two chemists working for Standard, Robert M. Thomas and William J. Sparks, had the idea of adding a small amount of butadiene to the polymer, so it could be vulcanized like rubber. Isoprene was later found to be better than butadiene. In contrast to most synthetic rubbers, butyl—as the new material was called—was polymerized at −150°F using a strong acid (boron triflouride) as the catalyst. The first large-scale experiments were carried out on a rooftop using a simple washing machine as the reactor and liquid ethylene as the coolant.[111]

At first sight, butyl appeared to be an outstanding synthetic rubber. It had a very low gas permeability and good aging resistance. Furthermore, the monomer was a cheap, readily available refinery by-product, and the polymerization process was simple. By August 1942, the RRC had authorized 132,000 long tons/year of butyl capacity (compared with five times that amount of GR-S capacity).[112] However, several problems arose. While butyl is an excellent rubber for inner tubes, it was a poor tire carcass material. It was also found to be incompatible with natural rubber. Furthermore, there were problems getting the plants into operation. Jersey Standard and its affiliate, Humble Oil, produced only 48,000 long tons of butyl in 1945 at their plants in Baton Rouge, Louisiana, and Baytown, Texas. Butyl was also produced by the Polymer Corporation at Sarnia, Ontario.[113]

Butyl's incompatibility with natural rubber or the copolymers became serious when the tubeless tire began to replace inner tubes. In 1950, Robert Morrissey (at Goodrich) decided to solve the problem by brominating the polymer. This was carried out simply by mixing the rubber and bromine in a Banbury mixer, a frightful but successful procedure. This bromobutyl was introduced as Hycar 2202 in 1954. It was not a commercial success, and Goodrich sold the technology to Polymer in 1965. Polymer relaunched bromobutyl in 1970.[114]

Morrissey had also looked at the chlorination of butyl, with some success. It was Jersey Standard, however, that commercialized chlorobutyl. Robert Thomas and Francis P. Baldwin discovered that gaseous chlorination could degrade the polymer and therefore switched to chlorination in solution. Standard introduced chlorobutyl in 1960.[115] Shortly before this, in 1959, Standard brought out a butyl passenger tire, which promised a "softer, quieter rid[e]."[116] Alas, the earlier doubts about butyl tires were well-founded. The new tires wore badly and were soon withdrawn.

Success or Failure?

The rubber research program was a wartime emergency effort that became a peacetime research program. The wartime effort achieved most of its goals; the peacetime program was markedly less successful. During the war, patriotism and the desire not to let down the boys at the front acted as powerful stimuli. Once the war was over, such psychological motivation was considerably reduced. The rubber companies once again gave priority to profits, technological advantage, and market position. There was no advantage (even a positive disadvantage) in making technical breakthroughs that would have to be shared with competitors. The universities began to see the research program as an easy means of funding their research and graduate students. While they clearly tried to produce worthwhile results, in order to create a rationale to prolong the research program, the academic groups nevertheless selected topics which reflected their own research interests.

Commercial competition probably provides the best spur to innovation in peacetime.[117] The wartime industry was made possible by the research performed by three competing groups (Goodrich-Phillips, Goodyear-Dow-Shell, and Jersey Standard-Firestone-U.S. Rubber) in the six years before Pearl Harbor. Companies outside or on the fringes of the rubber research program played an important role in the postwar development of synthetic rubber. Phillips Petroleum is noteworthy for its work on carbon blacks, tertiary mercaptans, cold rubber, and also stereoregular polybutadiene. General Tire not only developed oil-extended rubber but also assisted Charles Price, then at Notre Dame University, with the development of polypropylene oxide rubber. The Polymer Corporation in Canada commercialized a number of developments, including an improved cold rubber recipe, oil-extended rubber (with General Tire), latex-blended high-styrene resins (1949), synthetic Balata (*trans*-polyisoprene; 1964), and bromobutyl. Hercules introduced disproportionated rosin soap as an emulsifier and cumene hydroperoxide as an initiator; in the early 1960s, it contributed to the development of the important ethylene-propylene terpolymers. Du Pont, which had refused to join the rubber research program, developed fluorocarbon rubbers and other specialties in the 1950s.[118] Furthermore, the burst of innovation associated with synthetic "natural rubber" was associated with the revival of the competitive spirit that followed the decision by Congress to return the industry to the private sector.

The marketplace can be replaced by the peer-review system used by pure science or by mission-directed research.[119] Neither method was adequately employed by the postwar research program. Shortly after Williams's departure in mid-1943, the research program became an essentially self-governing program: each group did what it wanted to do, subject only to limited oversight.[120] This development inhibited the establishment of a rigorous review procedure. In contrast to the proposal-refereeing-approval cycle now used to monitor the funding of pure science, there was no system to review current research. The Polymer Research Discussion Group was an excellent forum to discuss and criticize recently completed work. The section personnel made visits to the laboratories and read the incoming reports. Nonetheless, there was no formal peer review system to assess the completed work or to give firm guidance about future research.

The narrowly defined wartime mission—the production of an effective synthetic rubber in the shortest possible time—was not replaced by specific postwar goals. A coherent overall strategy was lacking after 1945, and the program simply wandered. It was the absence of central planning that provoked Robert Solo to describe the research program as "leaderless and directionless."[121] Most of the useful research, especially by the academic groups, stemmed from the recommendations in R. R. Williams and Calvin Fuller's key 1942 memorandum. Ironically the very clarity of the program's wartime mission may have contributed to the failure to formulate a long-term strategy. Rather than impose a course that might be wrong, the central administration was inclined to let the research groups chose their own direction, trusting that they would see the best way forward.

The program's postwar mission was clear only to the rubber companies, which would have pursued similar goals in the absence of the rubber research program. The universities were different. They did not have boards of directors or even research directors. They were not at the cutting edge of rubber technology nor were they in continual contact with rubber consumers. In short, the universities were not in a position to see the whole picture for themselves. Given this situation, the research section could have done more to inform the leading academics about the overall situation, and to suggest goals that were within the scope of the universities but would have still made a significant contribution to the development of synthetic rubber.

The academic research groups did not extend the boundaries of polymer science very much because they conscientiously labored, for the most part, on applied problems. At the same time, however, they often failed to make a major contribution to the solution of these problems. This failure was probably a reflection of their inexperience in rubber technology. The universities also found it difficult to bridge the gap between basic research and practical application. They did not receive any significant assistance in overcoming this problem from the central program administration or the rubber companies. The Government Laboratories were of some use in this respect—for example, the collaborative research on new rubbers with the University of Illinois—but this unique institution was not exploited as fully as it could have been.

The absence of clear peacetime goals stemmed in part from a lack of strong leadership. The program had begun well, with gifted leaders with considerable expertise in chemical research at the helm: Bradley Dewey, Ray Dinsmore of Goodyear, Edwin Gilliland of the Massachusetts Institute of Technology, and R. R. Williams.[122] Unfortunately, they soon left to exercise their prodigious talents elsewhere. The departure of R. R. Williams was a grievous loss to the program. With his experience of directing research at America's premier research organization, his knowledge of rubber technology, and his work on vitamin B1, Williams was uniquely qualified to guide the rubber research program in peacetime. Successful government programs have often had strong charismatic leaders, such as Leslie Groves (atomic bomb) and Hyman Rickover (nuclear submarine). These men led from the front; provided clear goals; and carried their programs through periods of financial, technical, and political problems. The Rubber Directors, Jeffers and Dewey, were in this mould, but they were not replaced by men of equal stature.

The postwar program was also weakened by a lack of consistent support from Congress. The political response to a crisis often goes through a cycle. Congress is slow to act on warnings of a pending crisis, but when the crisis breaks there is an emphatic legislative response, in this case the Gillette bill. As the crisis eventually ebbs, the enthusiasm of Congress for the emergency program it has created wanes. Disillusionment sets in and the program is finally abolished. This cycle is true for the rubber program, the Reconstruction Finance Corporation, the spaceflight program, and above all, the synthetic fuel program.[123] In addition to the usual postcrisis decline in support, the synthetic rubber program also suffered from the attacks of the House Republicans, who won a majority in the 1946 elections.

This lack of political support was reflected in the underfunding of the research program. The total cost of the research program between 1943 and 1955 was $55.6 million, in contrast to a total capital investment in 1945 of $677 million.[124] It was less than half the amount the Massachusetts Institute of Technology alone received from the wartime Office of Scientific Research and Development over a much shorter period of time.[125] This expenditure was 1.9% of the value of GR-S produced during the program ("turnover"), and since the price of GR-S was held artificially low for several years, the true percentage must be even lower. The rubber industry traditionally spent less on research and development than the chemical industry. Goodrich only spent an average of 1.9% of sales on research and development between 1941 and 1954,[126] compared with the 3% of sales laid out by Du Pont during the 1930s.[127] Nonetheless, the government grants to Goodrich up to June 1952 ($2.5 million) were only 3% of Goodrich's total research and development spending ($86 million between 1943 and 1952). If the rubber program's research and development expenditure had been pegged at 3%, to allow for the fixed price of synthetic rubber—which had the effect of deflating the value of the turnover—it would have increased by over $2.5 million a year.

These problems could have been solved by returning the synthetic rubber research to the private sector immediately after the war, which was the original intention of the program's leaders.[128] Why was the research program extended beyond 1946 or 1947, when it was planned as a purely wartime emergency operation? Bureaucratic empire-building and the continuing search for a better synthetic rubber are two possible answers. My own research indicates, however, that the survival of the rubber program was a consequence of the Cold War. A native uprising in the Dutch East Indies and Communist bandit attacks in Malaya indicated that there was a real threat of the rubber-producing areas falling into anarchy or even under Communist control before the rubber plantations came back into full operation. The danger of World War III was ever present in a period that witnessed the Soviet takeover of Czechoslovakia, the Berlin airlift, and the Greek Civil War. There was still a pressing need to find an adequate substitute for natural rubber. The possibility of war with the Soviet Union fueled the search for a low-temperature cold-resistant rubber that could withstand the Russian winters. Atomic energy and missile development are the two technologies most closely associated with the Cold War. The postwar synthetic rubber research was also driven by the need to maintain America's military prowess after World War II. An excellent study could be made of the development of synthetic rubber in the context of the Cold War, for example, the important contribution made by the armed forces' research branches in the late 1940s and 1950s.

It is therefore striking that the rubber program did not seek to produce the specialty rubbers required by the armed forces. The low-temperature-resistant and oil-resistant rubbers were developed by the armed forces' own research branches. While this was probably an efficient division of labor, it weakened the political case for the existence of the rubber program. Once cold rubber had been successfully introduced in 1949, the program's leaders found it increasingly difficult to justify continued funding by the Federal Government. The rapid expansion of the armed forces' direct funding of basic research in the late 1940s and early 1950s also undercut the program's rationale as a patron of fundamental polymer research. For example, the Air Force displaced the rubber program as the leading sponsor of Carl Marvel's research in the 1950s.

The perceptive reader will notice that my conclusions are not dissimilar to those reached by Robert Solo.[129] The similarity of our conclusions are all the more remarkable given our different sources and methodologies. Nevertheless, while we may agree on several points, we differ about the solution to these failings. Solo argued the case for a strong central authority answerable for the rubber research program's success or failure. While I do not underestimate the value of strong leadership (and regard its absence as one of the program's failings), I do not believe that Solo's proposed solution would have been feasible.

The powerful commercial interests involved in the development of synthetic rubber would have made all-powerful central control impossible, at least in peacetime. Large companies are simply unwilling to forego control over their operations (especially research and development) to the extent necessary for such a central organ to be effective. Furthermore, it would have been extremely difficult for an American government to justify such stringent control of industrial affairs in peacetime. The federal synthetic rubber program was acceptable to the rubber industry only if the rubber companies were given a free hand with their own research, and indeed a considerable degree of control over the program as a whole. In the absence of this understanding—and we have seen that Goodrich was dissatisfied with even this compromise—the government would have been forced to run its research program without the rubber companies or resort to direct state control of the rubber industry.

Solo's desire to have the central authority held fully responsible for the rubber research program's results is equally problematic. How could the outcome of the program be assessed? Solo considered it a failure, but most of the participants in the program believe that it was a great success. The Baruch Committee could have been reconvened for this purpose, but its members would have been placed in a very difficult position, especially Bernard Baruch, who was not technically qualified. A technical panel could have been assembled, for example, by the National Academy of Sciences, but if the recent debate over the Strategic Defense Initiative is any guide, it would have been quickly mired in intrigue and controversy. Even if we assume that a consensus about the outcome of the program could have been reached, what appropriate rewards or punishment could have be given to those held responsible?

I also disagree with Solo's assessment of the rubber research program's technological achievements as "zilch, zero, nothing."[130] This study has shown that the research program enjoyed a number of technical successes. It solved several major production problems, including the agitation problem, the induction period caused by the soap, and the formation of popcorn polymer in recycled butadiene. The program encouraged academic research groups to assist the development of cold rubber. The academic groups were responsible for several low-temperature recipes, including the diazothioethers, an early version of the "Custom recipe," the Veroxasulfide recipe, and the peroxamine recipe. The government program also pioneered oil-extended rubber independently of General Tire and laid the foundations for the development of synthetic "natural rubber." I believe that the research program could have done more, perhaps much more, to accelerate innovation, but it was not by any means a complete failure. The wartime program was remarkably successful, given the difficult conditions under which it labored. The postwar program was less fruitful, but still made a notable contribution to the development of synthetic rubber.

This discrepancy in our findings arises because Solo fails to distinguish between the various innovations made during the program. The improvement of the original GR-S, the development of cold rubber and oil-extended rubber, and the development of an all-*cis* polyisoprene represent different types of innovation. The gradual improvement of the original polymer was incremental innovation. Lithium polymerization and the Ziegler catalysts, new methods of polymerization that were a quantum leap from the previous processes, were radical innovations. The other two innovations fell between these extremes. Cold rubber was a butadiene-styrene copolymer, like the original GR-S, but the new low-temperature recipes were markedly different from the mutual recipe. The addition of oil to rubber was not a wholly new idea, but it was a break with existing practice to use it to extend GR-S. These two innovations might be described as "semi-radical." These three types of innovation differ in several respects: time required to bring the innovation to maturity, cost of development, degree of originality, and ease of imitation. Using the rubber program as our source, we can construct a table to illustrate these differences (see Table 1).

Incremental innovation is both common and often underestimated.[131] It can be a change in the concentration of one component in a reaction mixture, the redesigning of a valve, or the introduction of a new test of the quality of the product. The usual aim is a slightly better process or product, and/or lower production costs. Incremental innovation is often associated with the removal of problems or bottlenecks in production. The soap problem and the agitation of the emulsion were two important examples of incremental innovation in the rubber program.

These two innovations occurred in the first year of the program and were successful. By 1945, the program had produced a significant improvement in the quality of GR-S and streamlined its production. This was a result of the advanced technological capabilities of the American rubber industry, patriotic desire during wartime to produce the best possible rubber, and rivalry between the plants with a view to being kept on after the war. It is also true, however, that the rubber program was well placed to assist incremental innovation, which is easy to manage, cheap to fund, and not very original, although it can display considerable ingenuity. Furthermore, the rubber companies were not too concerned about the exchange of information in this field. The advantages gained from such information, over the anticipated short time span, were small. The "tricks" that worked well in one plant might not

TABLE 1. The differences between the three types of innovation found in the American rubber program

Incremental	Semi-radical	Radical
Small technical improvements	Large technical improvement	Entirely new method
Generally low costs	Costs can be high	Costs can be very high
Time span of a few months to a year	Time span of two to five years	Time span of more than five years
Not always easy to imitate: "hands-on" experience often needed	Basic features can be easy to imitate	Key new features often easy to imitate
Originality is often low	Originality is high but it is not a conceptual breakthrough	Conceptual or technical breakthrough
Results in lower cost or better product	Usually a much better product or significant savings in costs	Can give innovator large profits and/or technical advantages
The advantage gained over other companies is small	Considerable advantage obtained over other firms	Marked advantage obtained over other firms, can result in dominance of industry

be so useful at another plant. This was an area in which it was possible to supply information without giving away the essential "know-how" needed to put these improvements into practice; that could only be achieved by an exchange of personnel, which rarely occurred.

The cost and time required for the development of cold rubber was higher than for most incremental innovations. It might be thought that the government-funded program was tailor-made for this type of innovation. It needed a significant amount of funding, wide-ranging experimentation with catalyst systems and polymerization temperatures, and extensive road testing. At the same time, it was a logical extension of earlier work and did not require a major conceptual or technological breakthrough. It also might be thought that the synthetic rubber producers would have been motivated by the need to compete with natural rubber after the war. Yet this line of development was only taken up in earnest following the shocking discovery that Germany was ahead in this field (compare the Sputnik "scare" of 1957). Furthermore, much of the industrial research that prodded the Reconstruction Finance Corporation into adopting the new process was carried out by the "outsider" Phillips Petroleum.

Two underlying reasons can be given for the delay in the development of cold rubber. As the big four rubber companies were tire manufacturers first and synthetic rubber producers only second, their desire to use natural rubber in tires more than counterbalanced their incentive to develop cold rubber. Furthermore, if the whole industry switched to cold rubber, as permitted by the cross-licensing agreements, no single firm would benefit. It is therefore not surprising that the larger companies did not seek to accelerate its development.

Research that focuses on the creation of radical innovations usually takes longer to produce results. While radical innovations, because of their novelty, often take longer than incremental innovations to bring to the market, it is the long gestation or induction period before a successful radical innovation appears on the laboratory bench or drawing board that characterizes this area of research as "long-term" research. Furthermore, only a few radical innovations succeed in the marketplace. The development of a radical innovation is literally a gamble on a very large scale. It is the large financial and strategic gains of developing a successful radical innovation that makes the gamble worthwhile.

Individuals and small companies may be motivated by the desire for wealth or even the desire to be socially useful. Large established companies, on the other hand, are forced to be innovative by competition. Innovation is disruptive in a mature bureaucratic organization, and would be avoided if it were possible, but companies operating in a science- or technology-based industry have little choice.[132] The financial and organizational costs of pursuing new technologies may be high, but a firm that did not innovate in a competitive industry would sooner or later go the way of sedan-chair carriers and oil-lamp makers. Radical innovations can displace decades of incremental innovation. The displacement of the Welsbach gas mantle by electric lighting and the propeller-driven aircraft by the jet plane are two prominent examples.

If the large rubber companies were reluctant to develop cold rubber within the government rubber program, this was even more the case for all-*cis* polyisoprene. The attempt by Firestone's top management to delay the development and then the announcement of Stavely's work can be seen to fit into the same pattern. It will also be recalled that Goodrich entrusted the development of a synthetic "natural rubber" to its "private" American Rubber Team, and then to the "external" Goodrich-Gulf. The disincentives for sharing information about radical innovations are considerable. The essential elements of radical innovations are often easy to communicate: use lithium instead of sodium, make a catalyst from aluminum triethyl and titanium tetrachloride. When several companies know the basis of an innovation (and it is not covered by an exclusive patent), the large profits and the strong technical position otherwise conferred by it disappear. The heavy costs and long development period of radical innovations are unsupportable without the hopes of profits and strategic advantage.

It is thus evident that government funding and cooperative research will only rarely be the best means of stimulating innovation, especially radical innovation. Furthermore, it appears that the government is often unsuccessful at promoting innovation in the private sector. It is significant that the three successful government-sponsored technological projects mentioned by Solo[133]—Manhattan Project, the nuclear submarine program, and NASA—were not in areas of prior commercial activity. By contrast, the only failure he cites (the housing research program sponsored by George Romney, Nixon's Secretary of Housing and Urban Development) was in an area dominated by existing companies. This is also true of most energy projects, including the ill-fated synthetic fuels program. This reinforces my belief that vigorous competition is the most effective means of promoting innovation in sectors characterized by large pre-existing firms.[134]

Notes

1. Of the eighteen Ph.D.'s known to be working for U.S. Rubber in 1943, at least nine had Ph.D.'s in physical chemistry or physics. There were only three who stated that they had Ph.D.'s in organic chemistry. The information for these statistics was taken from various editions of *American Men of Science*.

2. Of the twenty Ph.D.'s known to be working for Goodrich in 1943, at least ten had Ph.D.'s in organic chemistry, compared with five for physical chemistry and none in physics.

3. Interview of John McCool by Peter Morris, 6 November 1986.

4. Beatty interview. Also see W. C. Warner, "Arthur Edgar Juve," in Wyndham D. Miles (ed.), *American Chemists and Chemical Engineers* (Washington, D.C., 1976).

5. Of the fourteen Ph.D.'s known to be working for Goodyear in 1943, four had Ph.D.'s in organic chemistry and five in physical chemistry, colloid chemistry, and physics.

6. Interview of Robert Pierson by Peter Morris, 6 November 1986.

7. Interview of Harold P. Brown by Colleen Wickey, 12 March 1986. Also see McCool interview.

8. McCool interview.

9. Annual Reports of B. F. Goodrich, 1947 (p. 5), 1948 (pp. 4–5), and 1949 (p. 4).

10. O'Callaghan, *The Government's Rubber Projects*, volume 2, 546–547. Solo, "Research and Development," 69. Private Communication from P. S. Greer to Peter Morris, 9 February 1987.

11. Solo, *Synthetic Rubber*, 103. Solo implies that both General and Phillips were brought into the research program, but this is incorrect. General never had a research agreement with the RFC, only a development agreement as an operator of a GR-S plant. Private communication from J. D. D'Ianni, 3 August 1987.

12. Rowzee and Buckler interview.

13. Howard, *Buna Rubber,* 162

14. For a brief historical overview of both American and German polymerization methods, see R. L. Bebb, E. L. Carr, and L. B. Wakefield, "Synthetic Rubber Polymerization Practices," *Industrial and Engineering Chemistry* 44 (1952), 724–730. For a more detailed account of the development of emulsion polymerization of GR-S (including cold rubber), see Charles F. Fryling, "Emulsion Polymerization Systems," in Whitby, Davis, and Dunbrook (eds.), *Synthetic Rubber,* 224–283; for the origins of the mutual recipe see page 243. Interview of Ben Kastein by Peter Morris, 21 May 1986.

15. C. F. Fryling and B. M. G. Zwicker, "The Effect of Rate of Stirring Upon the Properties of GR-S," CR 19, 23 March 1943. B. M. G. Zwicker, "Effect of Variable Agitation on GR-S Polymerization," CPD 95, 28 June 1943.

16. B. F. Goodrich Co., "Controlled Viscosity Polymers," CPD 156, 29 October 1943.

17. J. Lawrence, R. W. Hobson, and A. M. Borders, "Effect of Continuous Modifier Addition on Intrinsic Viscosity and Approximate Weight Distribution," CR 212, 6/7 December 1943.

18. I. M. Kolthoff and W. E. Harris, *Industrial and Engineering Chemistry, Analytical Edition* 18 (1946), 161–162.

19. There is an extensive literature on the kinetics of modification. For two contemporary summaries, see P. V. Smith, Jr., "Mercaptans as Modifiers in the Copolymerization of Butadiene and Styrene" (University of Illinois Ph.D. dissertation, part II, 1945), 38–39; and R. F. Dunbrook, *India Rubber World* 117 (1947), 355–359. Also see the interview of Elbert E. Gruber by Peter Morris, 5 November 1986, for a short account of the wartime research. I. M. Kolthoff and W. E. Harris, "Report XXVII: A Comparison between Primary and Tertiary Mercaptans as Modifiers and Activators in the Mutual Recipe," CR 355, 12/13 June 1944. Harris and Kolthoff, "Report LII: A Comprehensive Review of Modification, Disappearance Curves of and Activation by Mercaptans in the GR-S Recipe," CR 626, 23 February 1945. Kolthoff and Harris, *Journal of Polymer Science* 2 (1947), 49–71. W. B. Reynolds and P. J. Canterino "Oxidation of Mercaptans in Soap Micelles—The Diffox Product," CR 1073, 25 April 1946. R. L. Frank, P. V. Smith, F. E. Woodward, W. B. Reynolds, and P. J. Canterino, *Journal of Polymer Science* 3 (1948), 39–49.

20. See, for example, R. L. Frank, J. Blegen, P. V. Smith, and H. Ward, "Investigation of Phillips Sulfole B–8 Mercaptans," CR 341, 9 June 1944.

21. The sulfole tertiary mercaptans are first mentioned in the monthly synthetic rubber progress reports in the "Progress Reports" in the B. F. Goodrich archives, University of Akron archives, in March 1943.

22. R. F. Dunbrook, *India Rubber World* 117 (1947), 355.

23. Livingston and Cox, "The Manufacture of GR-S," 190; Pierson interview. Also see R. C. Reed, "Polymerization Soap Review," CPD 149, 21 October 1943.

24. Traced in the monthly reports to the Office of the Rubber Director (ORD), for February 1944 (CR 298), March 1944 (CR 313), April 1944 (CR 330), May 1944 (CR 367), and June 1944 (CR 389). Also see C. S. Marvel, W. E. Blackburn, D. A. Shepherd, R. J. Dearborn, and J. A. Dammann, "Investigation of Soaps in GR-S Polymerization," CR 273, 22 March 1944; "Investigation of Soaps in GR-S Polymerization II," CR 391, 27 July 1944. "The University's Synthetic Rubber Program During World War II," November 1956, UI 39/1/8 Box 1, 6–7.

25. Livingston and Cox, "The Manufacture of GR-S," 189.

26. Pierson interview. Fryling, "Emulsion Polymerization Systems," 251–252.

27. *Ibid.*, 262–263.

28. Eli M. Dannenberg, "The Carbon Black Industry: Over a Century of Progress," in *Rubber Division 75th Anniversary* (Akron, 1984), 35–40. R. P. Dinsmore and R. D. Juve, "The Processing and Compounding of GR-S," in Whitby, Davis, and Dunbrook (eds.), *Synthetic Rubber*, 399–406. Beatty interview.

29. Dannenberg, "The Carbon Black Industry," 40.

30. C. F. Fryling, R. L. Bebb, L. H. Howland, and J. D. D'Ianni, "Report of the Committee on Revision of the GR-S Formula to Include a Catalyst Activator," CR 116, 12 August 1943, 3.

31. P. S. Greer, "Summary Report X. Effect of Temperature in the Polymerization of Buna S," CPD 79, 8 June 1943. W. E. Messer and L. H. Howland, "Influence of Variation on Polymerization Formula, Polymerization Temperature, Conversion, and Amount of Carbon Black Upon Various Properties of GR-S," CR 91, 24/25 June 1943, Table B. Also see the review by L. W. Howland and M. W. Swaney, "Influence of Temperature of Reaction on the Properties of Buna S Polymers," CR 142, 13/14 September 1943.

32. Fryling, "Emulsion Polymerization Systems," 261. The Goodrich recipe became known as the "H Formula" and was studied by several groups, including U.S. Rubber at Naugatuck and Kharasch's group at the University of Chicago. See, for example, M. S. Kharasch, "Copolymerization at 30°C. I. The H Formula," CR 353, 12/13 June 1944. For a brief history of cold rubber up to 1948, see W. H. Shearon, Jr., J. P. McKenzie, and M. E. Samuels, "Low Temperature Manufacture of Chemical Rubber," *Industrial and Engineering Chemistry* 40 (1948), 769–777.

33. Gruber interview. C. F. Fryling and W. D. Stewart, "Catalyst, Promoter and Modifier Systems For Use With the GR-S Type of Recipe," CR 20, 23 March 1943.

34. Messer and Howland, CR 91, Table E.

35. Fryling, "Emulsion Polymerization Systems," 261.

36. Pierson interview.

37. H. W. Starkweather, et al., *Industrial and Engineering Chemistry* 39 (1947), 210–212.

38. Fryling, "Emulsion Polymerization Systems," 270.

39. J. E. Troyan, "Report of the Copolymer Process Development Branch" to the 13th meeting of the Rubber Research Board, 12 January 1944, RG 234, RRC, Entry 231, PI–173, Minutes of the Rubber Research Board, 6.

40. J. C. Elgin, "Report of the Copolymer Development Branch" to the 14th meeting of the Rubber Research Board, 9 February 1944, 4. A. S. Gow and P. S. Greer, "Progress Report: GR-S Polymerization Activation," CD 33, 22 February 1944.

41. "Summary of . . . Meeting No. 15," 8 March 1944, 6.

42. J. C. Elgin, "Report of the Copolymer Development Branch" to the 15th meeting of the Rubber Research Board, 4.

43. "Summary of . . . Meeting No. 18," 26 July 1944, 2.

44. One of the first reports was J. W. Livingston, "Synthetic Rubber Research at Leverkusen, Germany," a report of the Rubber Subcommittee of the Technical Industrial Intelligence Committee, 27 September 1945; published by the Department of Commerce as PB 13356 (available at the Library of Congress) and as CR 810/CD 458. R. L. Bebb and L. B. Wakefield, "German Synthetic-Rubber Developments," in Whitby, Davis, and Dunbrook (eds.), *Synthetic Rubber*, 971–981. E. Konrad and W. Becker, "Zur Geschichte des bei tiefer Temperatur polymerisierten synthetischen Kautschuks," *Angewandte Chemie* 62 (1950), 423–426.

45. I. M. Kolthoff and E. J. Meehan, "Progress Report—April 1946," CR 1092, 82–89; Kolthoff and Meehan, "Progress Report—July 1946," CR 1176, 56–57. Kolthoff, Meehan, and J. A. Cain, "Report XCIII: Butadiene-Styrene Copolymerizations in Redox

Recipes. Part I. Alkaline Recipes with S. F. Flakes," CR 1251, 7 October 1946; Kolthoff, Meehan, and R. Held, "Report XCIV: Butadiene-Styrene Copolymerizations in Redox Recipes. Part II. Alkaline Recipes with Rosin Soap," 7 October 1946. Kolthoff and A. I. Medalia, *Journal of Polymer Science* 5 (1950), 391–427. For a general account of this type of recipe, see F. A. Bovey, I. M. Kolthoff, A. I. Medalia, and E. J. Meehan, *Emulsion Polymerization* (New York, 1955), 377–390.

46. G. S. Whitby, "Introduction" to Whitby, Davis, and Dunbrook (eds.), *Synthetic Rubber*, 14–15. E. J. Vandenberg and G. E. Hulse, *Industrial and Engineering Chemistry* 40 (1948), 932–937.

47. C. F. Fryling and W. M. St. John, "Emulsion Polymerization of Synthetic Rubber With a Cumene Hydroperoxide-Redox Recipe. Part II. The Effect of Temperature of Polymerization on the Physical Properties of Rosin Rubber," CR 1403, 12 February 1947.

48. O'Callaghan, *The Government's Rubber Projects*, volume 2, 574–577. "Cold Rubber," Memorandum RDDR-150 from P. S. Greer to E. D. Kelly, 17 February 1954, a copy supplied by Mr. Greer is available at the Beckman Center for History of Chemistry.

49. O'Callaghan, *The Government's Rubber Projects*, volume 2, 576.

50. Greer, "Cold Rubber," 1–2.

51. Calculated from an Office of Synthetic Rubber production chart, 8 December 1955, supplied by P. S. Greer and stored at the Beckman Center.

52. "Die Emulsion-Polymerisation des Butadiens im Gemisch mit anderen polymerisiebaren Stoffen. (Mischpolymerisation)," a paper presented to the I.G. Farben Scientific Rubber Conference, June 24/25 1930, folder 153/3.2, Bayerwerkarchiv, Bayer AG, Leverkusen, Germany.

53. Logemann and Pampus, "Buna S," 482.

54. E. Weinbrenner, "Zur Entwicklung von Buna S3 und Buna S4," a paper presented to the 11th meeting of I.G. Farben's Rubber Technology Committee (Kauteko), 16 June 1944, FD 1502/48 contained within FD 3452/46, Imperial War Museum, London, England.

55. L. E. Ludwig, D. V. Sarbach, B. S. Garvey, Jr., and A. E. Juve, *India Rubber World* 111 (1944), 180–186.

56. L. E. Ludwig, D. V. Sarbach, B. S. Garvey, Jr., and A. E. Juve, *India Rubber World* 112 (1945), 731–737.

57. Solo, *Synthetic Rubber*, 101–102. Gruber interview. Brown interview.

58. O'Callaghan, *The Government's Rubber Projects*, volume 2, 578–579. Greer interview.

59. "New Oil-Extended Rubber has Commercial Possibilities," *Chemical and Engineering News* 29 (1951), 5058. For the subsequent legal activities, see Greer interview; and Herbert and Bisio, *Synthetic Rubber*, 153–155.

60. D'Ianni, "Fun and Frustrations with Synthetic Rubber," G73.

61. J. W. Adams and L. H. Howland, "Latex Masterbatching," in Whitby, Davis, and Dunbrook (eds.), *Synthetic Rubber*, 668.

62. A. Beck and M. Müller-Conradi, U.S. Patent 1,991,367, filed (in Germany) 24 April 1930, (in U.S.A.) 13 April 1931.

63. Adams and Howland, "Latex Masterbatching," 678.

64. For the early experimentation wtih carbon black masterbatching, see "Summary of . . . Meeting No. 15," 8 March 1944, RG 234, RRC, Entry 231, PI–173, Minutes of the Rubber Research Board, 11.

65. Adams and Howland, "Latex Masterbatching," 678.

66. *Ibid.*, 678–679.

67. *Rubber: Production, Shipments and Stocks,* U.S. Department of Commerce, Bureau of the Census, Current Industrial Reports, MA 30A(84)–1 (1984), 2.

68. W. B. Reynolds, "GR-S Polymerization: Diazo Thioethers as Catalyst-Modifiers," CR 408, 4 June 1944. Reynolds and E. W. Cotten, *Industrial and Engineering Chemistry* 42 (1950), 1905–1910. Reynolds and Cotten, U.S. Patent 2,501,692, filed 17 January 1946. I. M. Kolthoff and W. J. Dale, "Report XLIX: p-Methoxy Phenyl Diazo Thio-(2-Naphthyl),-Ether (Denoted as M.D.N.), as Activator-Modifier in Polymerizations by Various Recipes," CR 522, 27 December 1944. Kolthoff and Dale, *Journal of Polymer Science* 3 (1948), 400–409. Kolthoff and Dale, *Journal of Polymer Science* 5 (1950), 301–306.

69. E. J. Buckler, "Canadian Contributions to Synthetic Rubber Technology," *Canadian Journal of Chemical Engineering* 62 (1984), 4–5. Rowzee and Buckler interview.

70. Solo, "Research and Development," 77–78; Solo, *Synthetic Rubber,* 101–102.

71. R. W. Laundrie, E. E. Rowland, A. N. Snyder, W. K. Taft, and G. J. Tiger, *Industrial and Engineering Chemistry* 42 (1950), 1439–1442.

72. Yakov M. Rabkin, "Technological Innovation in Science: The Adoption of Infrared Spectroscopy by Chemists," *Isis* 78 (1987), 43–45.

73. The development of infra-red spectroscopy in the petrochemical industry is well covered in the rubber program's RM reports. Also see R. Brattain and O. Beeck, *Journal of Applied Physics* 13 (1942), 699–705. Interview of Arnold O. Beckman by Jeffrey Sturchio and Arnold Thackray for the Beckman Center oral history program, 23 July 1985. The minutes of the summer school were circulated as RM–91.

74. Interview of Samuel Gehman by Peter Morris, 6 November 1986.

75. R. Bowling Barnes, Urner Liddel, and V. Z. Williams, *Industrial and Engineering Chemistry, Analytical Edition* 15 (1943), 83–90.

76. J. U. White and P. J. Flory, "Infra-Red Spectra and Structure of Synthetic Rubbers," CR 14, 20 February 1943. It was recirculated in April 1944 as General Report 10.

77. J. E. Field, D. E. Woodford, and S. D. Gehman, *Journal of Applied Physics* 17 (1946), 386–392. A similar analysis, with the same misindentification of the 967 cm^{-1} band, was carried out by Karl Luft of I.G Farben's Ludwigshafen laboratory in 1944, FD 3450/46, Imperial War Museum.

78. E. J. Hart and A. W. Meyer, *Journal of the American Chemical Society* 71 (1949), 1980–1985.

79. R. R. Hampton, *Analytical Chemistry* 21 (1949), 923–926.

80. Goodrich's Physical Research Laboratory was unable to assess the *cis* content of a polybutadiene using infra-red spectrophotometry in August 1950. Progress report of the Physical Research Laboratory for August 1950, 1. All the Goodrich progress reports cited here are bound in the "Progress Digests" for 1950, and 1951 (two volumes), B. F. Goodrich Collection, Series H, University of Akron archives. Firestone also took some time to perfect its infra-red analyses of polybutadiene and especially polyisoprene. See the interview of Fred Foster by Frank McMillan, 4 September 1973; the transcript copy was kindly supplied by Frank McMillan.

81. Howard, *Buna Rubber,* 281; the committee was the Special Committee of the U.S. Senate Investigating the National Defense Program (the Truman Committee).

82. O'Callaghan, *The Government's Rubber Projects,* volume 2, 581–582.

83. Progress report of the American Rubber Team for July 1950, 1. Conversation between Peter Morris and J. Roger Beatty, 7 August 1987. Quote taken from Herman E. Schroeder, "Facets of Innovation," *Rubber Chemistry and Technology* 57 (1984), G90.

84. Progress report of the Pioneering Research Department for April 1951, 1.

85. *Ibid.,* 7.

86. Progress report of the American Rubber Team for August 1951, 3. Beatty conversation. For the problems with the evaluation of test results, see the progress report of the Pioneering Research Department for April 1951, 7; and the progress report of the Physical Research Laboratory for May 1951, 2.

87. E. R. Weidlein, Jr., "Synthetic Rubber Research in Germany," *Chemical and Engineering News* 24 (15 March 1946), 771.

88. C. S. Fuller, "Some Recent Contributions to Synthetic Rubber Research," *Bell System Technical Journal* 25 (1946), 380–383.

89. A list of features necessary for a low hysteresis rubber drawn up by Goodrich in March 1950 (progress report of the Pioneering Research Department for February 1950, 2) does not mention the all-*cis* configuration. This was added three months later in the progress report of the Pioneering Research Department for May 1950, 2.

90. C. S. Marvel and W. J. Bailey, "Sodium as a Polymerization Catalyst for Butadiene and Butadiene-Styrene Mixtures (Preliminary Report)," CR 272, 13/14 March 1944; Marvel and Bailey, "Sodium as a Polymerization Catalyst for Butadiene and Mixtures of Butadiene and Styrene II," CR 520, 11/12 January 1945. Also see C. H. Schroeder, "Evaluation of Sodium Polymerized GR-S From the University of Illinois," CR 417, 6 September 1944. Marvel, Bailey, and G. E. Inskeep, *Journal of Polymer Science* 1 (1946), 275–288. W. K. Taft and G. J. Tiger, "Diene Polymers and Copolymers Other Than GR-S and the Specialty Rubbers," in Whitby, Davis, and Dunbrook (eds.), *Synthetic Rubber,* 737–747. Interview of W. J. Bailey by James J. Bohning for the Beckman Center oral history program, 3 June 1986. Additional information from Professor Bailey, telephone conversation, 20 August 1987.

91. Progress report of the Pioneering Research Department for May and June 1951, 1. Quote taken from the progress report of the American Rubber Team for August 1951, 2.

92. A. A. Morton, E. E. Magat, and R. L. Letsinger, *Journal of the American Chemical Society* 69 (1947), 950–961. Morton, *Industrial and Engineering Chemistry* 42 (1950), 1488–1496. Morton, *Rubber Age* 72 (1953), 473–476. Taft and Tiger, "Diene Polymers," 748–752.

93. This conclusion is partly based on information supplied by Attilio Bisio, September 1987.

94. K. Ziegler, F. Dersch, and H. Wollthan, *Annalen der Chemie* 511 (1934), 13–44.

95. K. Ziegler, F. Häffner, and H. Grimm, *Annalen der Chemie* 528 (1937), 101–113.

96. K. Ziegler, H. Grimm, and R. Willer, *Annalen der Chemie* 542 (1940), 90–122.

97. T. Midgley, Jr., "Synthetic and Substitute Rubbers," in C. C. Davis and J. T. Blake (eds.), *Chemistry and Technology of Rubber* (New York, 1937), 683. Progress report of the American Rubber Team for August 1951, 2.

98. M. S. Kharasch and W. Nudenberg, "Preliminary Report on Polymerizations with Alkali Hydrides," CR 427, 14/15 September 1944.

99. C. Walling, "The Contributions of Morris S. Kharasch to Polymer Chemistry," in W. A. Waters (ed.), *Vistas in Free Radical Chemistry. In Memoriam, Dr. Morris S. Kharasch* (London, 1959), 146.

100. F. W. Stavely, et al., "Coral Rubber—A Cis-1,4,-Polyisoprene," *Industrial and Engineering Chemistry* 48 (1956), 778–783. Stavely, "Lithium Polymerization Catalysts," *Rubber Chemistry and Technology* 45 (1972), G56–G60; quote on page G58. Frank M. McMillan, *The Chain Straighteners,* 151–159, gives Fred Foster, a young research chemist in Benny Johnson's group, the credit for suggesting lithium polymerization. Foster

referred to Ziegler's paper on *cis*-2-butene in an internal memorandum of 11 July 1952, but ironically stressed the use of sodium, not lithium. It is possible Stavely was already considering the use of lithium. Foster in his interview by McMillan stated that Stavely was the first to think of replacing butadiene with isoprene. I am very grateful for Frank McMillan's unstinting assistance in this matter, including transcript copies of the relevant interviews and a photocopy of Foster's memorandum.

101. K. Ziegler, E. Holzkamp, H. Breil, and H. Martin, *Angewandte Chemie* 67 (1955), 541–547. McMillan, *The Chain Straighteners*, 56–72.

102. S. E. Horne, et al., "Ameripol SN—A Cis-1,4-Polyisoprene," *Industrial and Engineering Chemistry* 48 (1956), 784–791. The account given here is based on an interview of Carlin F. Gibbs by Peter Morris, 6 August 1987, and is compatible with Frank McMillan's interviews of Samuel Horne (6 September 1973), and Morton Golub (29 February 1972). Once again, I would like to thank Frank McMillan for his help in elucidating these events. Somewhat different accounts have been published by Horne and McMillan. Horne, "Polymerization of Diene Monomers by Ziegler Type Catalysts," *Rubber Chemistry and Technology* 53 (1980), G68–G79; Horne, "The History of Synthetic Rubber," 7–8. McMillan, *The Chain Straighteners*, 143–151.

103. "Synthesizing 'Natural' Rubber," *Chemical and Engineering News* 33 (1955), 4518. McMillan, *The Chain Straighteners*, 159–160.

104. Morris's interviews of Gibbs and Greer—and also Frank McMillan's interview of N. R. Legge of Shell Development Co. (8 October 1972)—agree on this point. McMillan (*The Chain Straighteners*, 151) states that Goodrich agreed, under legal pressure, to license the other companies.

105. George L. Clark, "Rubber, As It Is Revealed by X-Rays," *India Rubber World* 79 (1929), 55–59; reference to synthetic rubber on page 56.

106. Progress report of the Polymerization Research Department for May 1950, 2; and progress report of the Physical Research Laboratory for August 1950, 1.

107. Progress reports of the Technical Service Research Department for August 1950, 2; September 1951, 2; November 1951, 2.

108. Progress reports of the Technical Service Research Department for January 1951, 4; April 1951, 3; May 1951, 2.

109. "Man-made Polymers Akin to Natural," *Chemical and Engineering News* 32 (1954), 1451. W. J. Bailey, "The Synthesis of Oriented High Polymers," *Proceedings, Joint Army-Navy-Air Force Conference on Elastomer Research and Development* (Washington, D.C., 1954), 113–117. Bailey interview.

110. K. Ziegler and H-G. Gellert, *Annalen der Chemie* 567 (1950), 179–203. Ziegler and Gellert, *Angewandte Chemie* 64 (1952), 323–329. McMillan, *The Chain Straighteners*, 38–45.

111. For the history of butyl rubber, see: Howard, *Buna Rubber*, 47–58. Robert M. Thomas, "Early History of Butyl Rubber," *Rubber Chemistry and Technology* 42 (1969), G90–G96. Friedrich Holscher, *Kautschuk, Kunststoffe, Fasern* (BASF, 1972), 61–62. Tornqvist, "Polyolefin Elastomers," 144–148. Interview of Willard C. Asbury and A. Donald Green by Peter Morris for the Beckman Center oral history program, 9 December 1985.

112. Howard, *Buna Rubber*, 186. O'Callaghan, *The Government's Rubber Projects*, volume 2, 496–499.

113. R. F. Dunbrook, "Historical Review," 52.

114. Tornqvist, "Polyolefin Elastomers," 149. Buckler, "Canadian Contributions to Synthetic Rubber," 5–6. Gruber interview.

115. Tornqvist, "Polyolefin Elastomers," 149–150. F. P. Baldwin, "Modifications

of Low Functionality Elastomers," *Rubber Chemistry and Technology* 52 (1979), G79–G80.

116. Herbert and Bisio, *Synthetic Rubber,* 200.

117. For a critical discussion of this postulate, see R. R. Nelson and S. G. Winter, *An Evolutionary Theory of Economic Change* (Cambridge, Massachusetts, 1982), 357–365.

118. Schroeder, "Facets of Innovation," G86–G106.

119. Nelson and Winter, *Evolutionary Theory,* 391.

120. For the self-governing nature of the research program, see: G. S. Whitby, "Introduction," in Whitby, Davis and Dunbrook, *Synthetic Rubber,* 29. Greer interview; and Greer, "Rubber Program Notes." D'Ianni interview. For the similar situation in the early postwar Office of Naval Research, see David K. Allison, "U.S. Navy Research and Development," in Merritt Roe Smith (ed.), *Military Enterprise and Technological Change* (Cambridge, Massachusetts, 1985), 297; and Harvey M. Sapolsky, "Academic Science and the Military: The Years Since the Second World War," in Nathan Reingold (ed.), *The Sciences in the American Context: New Perspectives* (Washington, D.C., 1979), 385–386.

121. Solo, "Research and Development," 79.

122. R. P. Dinsmore, Goodyear's vice-president of research, was Assistant Deputy Rubber Director in charge of Research and Development for Synthetics until May 1943. He was replaced by Edwin R. Gilliland, who had developed the fluidized bed process for the catalytic cracking of petroleum with his famous MIT colleague Warren K. Lewis. Gilliland returned to MIT when the Office of the Rubber Director was dissolved on 1 September 1944. Williams returned to Bell Telephone Laboratories in May 1943. He left Bell Labs after the war to become director of grants at the Research Corporation. Fuller rejoined Bell Labs in June 1944. Nine years later, he made a key breakthrough in the development of silicon solar cells with Gerald Pearson. See Fuller interview; and Open University course T362: *Design and Innovation* (Milton Keynes, U.K., 1986), Block 1: "Invention," Unit 3: "Solar Cells," 60–62.

123. For the existence of this cycle in the case of the synthetic fuels program, see P. W. Hamlett, "Technological Policy Making in Congress: The Creation of the U.S. Synthetic Fuels Corporation," and M. M. Crow, "Synthetic Fuel Technology Nondevelopment and the Hiatus Effect: The Implications of Inconsistent Public Policy," in E. J. Yanarella and W. C. Green (eds.), *The Unfulfilled Promise of Synthetic Fuels: Technological Failure, Policy Immobilism or Commercial Illusion* (Westport, Connecticut, 1987).

124. Research funds calculated from an Office of Synthetic Rubber production chart, 8 December 1955, supplied by P. S. Greer and stored at the Beckman Center for History of Chemistry. Total capital investment in 1945 taken from Phillips, *Competition in the Synthetic Rubber Industry,* 43–44.

125. J. P. Baxter, *Scientists Against Time* (Boston, 1946), Appendix C, 456.

126. "Research and Development Costs vs. Sales," taken from Dr. Fritz's file on research costs, 6 July 1955, in "Research Costs and Contributions, 1943–1952 Inclusive," B. F. Goodrich Collection, Series H, Box 9, University of Akron archives. It is unclear if the total expenditure includes government grants. Goodrich's research and development spending steadily increased during the 1950s to 2.9% of sales in 1954.

127. W. J. Reader, *Imperial Chemical Industries: A History,* volume 2, *The First Quarter-Century, 1926–1952* (London: Oxford University Press, 1975), 94. Imperial Chemical Industries was lower at around 2.5%, *ibid.,* 94.

128. R. R. Williams, "Future Research on Synthetic Rubber," 30 June 1943, in casefile 24060–1, volume A, AT&T Archives, Warren, New Jersey. Private information from C. S. Fuller, 24 July 1987.

129. Solo, "Research and Development," 79–82; Solo, *Synthetic Rubber*, 108–109, 114.

130. *Ibid.*, vii–viii.

131. For a summary of the most important work in this area, see Patrick Kelly and Melvin Kranzberg (eds.), *Technological Innovation: A Critical Review of Current Knowledge* (San Francisco, 1978); U.S. Congress, Joint Economic Committee, *Research and Innovation: Developing a Dynamic Nation*, Special Study on Economic Change, volume three, 96th Cong., 2nd sess., 1980, Committee Print; and Nuala Swords-Isherwood, *The Process of Innovation* (British-North American Committee, 1984). Good introductions to the economics of innovation include Christopher Freeman, *An Introduction to the Economics of Industrial Innovation*, second edition (Cambridge, Massachusetts, 1982); Edwin Mansfield, *The Economics of Technological Change* (New York, 1968); Nathan Rosenberg, *Perspectives on Technology* (Cambridge, England, 1976); Nathan Rosenberg, *Inside the Black Box: Technology and Economics* (Cambridge, England, 1982); and Jacob Schmookler, *Invention and Economic Growth* (Cambridge, Massachusetts, 1966).

The best collections of case studies of innovation are still John Jewkes, David Sawers, and Richard Stillerman, *The Sources of Invention*, 2nd edition (New York, 1969), and J. Langrish, M. Gibbons, W. G. Evans, and F. R. Jevons, *Wealth from Knowledge: A Study of Innovation in Industry* (London, 1972).

Examples of the government's association with new technology that have been the subject of good historical analysis include the Manhattan Project, Project Apollo, and the Jet Propulsion Laboratory. See Richard G. Hewlett and Oscar E. Anderson, Jr., *A History of the United States Atomic Energy Commission*, volume 1, *The New World, 1939–1946* (College Park, Pennsylvania, 1962); John Logsdon, *The Decision to Go to the Moon, Project Apollo and the National Interest* (Cambridge, Massachusetts, 1970); Clayton R. Koppes, *JPL and the American Space Program: A History of the Jet Propulsion Laboratory* (New Haven, 1982).

For a pioneering study of the interactions between an oil company, its business competitors, and the government, see August W. Giebelhaus, *Business and Government in the Oil Industry: A Case Study of Sun Oil, 1876–1945*, Industrial Development and the Social Fabric 5 (Greenwich, Connecticut, 1980). The development of the petroleum refining industry was brilliantly analyzed in John L. Enos, *Petroleum, Progress and Profits: A History of Process Innovation* (Cambridge, Massachusetts, 1962).

There is a growing literature on corporate research. For recent examples, see David A. Hounshell and John K. Smith, *Science and Corporate Strategy: Du Pont R&D, 1902–1980* (New York, 1988); Stuart W. Leslie, *Boss Kettering* (New York, 1983); Leonard S. Reich, *The Making of American Industrial Research: Science and Business at GE and Bell, 1876–1926* (Cambridge, England, 1985); Jeffrey L. Sturchio, "Chemistry and Corporate Strategy at Du Pont," *Research Management* 27 (1984), 10–18; George Wise, *Willis R. Whitney and the Rise of American Industrial Research* (New York, 1985).

132. Donald A. Schon, *Technology and Change, The New Heraclitus*, (New York, 1967), 56–74. This is similar to Joseph Schumpeter's concept of "creative destruction," for which see Schumpeter, *Capitalism, Socialism, and Democracy*, 3rd edition (New York: Harper and Row, 1962), 81–86.

133. Solo, introduction to *Across the High Technology Threshold*, viii.

134. Compare Nelson and Winter, *Evolutionary Theory*, 392. Peter Spitz has written a good account of the deadening effect of the pre-World War II cartels on innovation, "The Old Order: Cooperation and Cartels," Chapter Six, *Petrochemicals*.

Chapter 3
The Universities

Introduction

Supporters of the program will protest that we have neglected key aspects of the university research in the last chapter, namely, an increased understanding of the underlying processes, and the development of new techniques. As we will see, this defense has merit, but were these contributions sufficient to make a government-sponsored research program worthwhile? Was the $7 million spent on the university research a good investment for the American taxpayer? Did government intervention actually lead to an increased contribution from the universities?

The first part of this chapter examines the main features of the research carried out at four leading universities in the program: Illinois, Chicago, Minnesota, and Cornell. These four institutions accounted for three-fifths of the grants given by the program to universities between 1942 and 1952.[1] Furthermore, they represent different aspects of the rubber research, as the section heads indicate. It is interesting that each group had a distinct character, which was largely a reflection of the prewar interests of the group leader. The role of the Polymer Research Discussion Group as a means of disseminating information is discussed in the context of three prewar forerunners.

Prior Industry-University Collaboration

The interwar period saw the rapid development of industrial support for fundamental research in the universities.[2] By the beginning of World War II, most of the senior chemists who later took part in the university-based research program were accustomed to working with and being sponsored by chemical companies. They were at ease discussing industrial problems with corporate scientists. They did research in their laboratories which served the double function of aiding industry and advancing pure chemistry. Industrial fellowships—grants provided by companies to support postgraduates and postdoctorals—were commonplace by the late 1930s, if not plentiful.

It should not be thought, however, that such cooperation was true across the board, or that the relationship was invariably a happy one. The collaboration benefitted, for the most part, a select group of research-oriented universities. Companies also tended to insist that the academic research they sponsored should be directed toward very specific problems, or if the research was of industry-wide applicability, that it should not be published in the open literature.

To what extent were these problems solved in the rubber research program? The program used the same select group of universities, just as the National Science Foundation did in the 1950s and 1960s. Indeed, as might be expected in view of the specialized nature of the program, it was even more selective. On the other hand, the emphasis on the free exchange of information removed the problem of publication vetoes, and external publication was strongly encouraged after 1945. Furthermore, an early decision by the Office of the Rubber Director to give the universities free rein liberated them (for better or worse) from the pressure to concentrate on problems of immediate interest to the industry.

Illinois: Chemical Structure and Organic Synthesis

The University of Illinois at Urbana was the most heavily funded of all the academic institutions in the synthetic rubber research program; it received, in total, almost 50% more than the second-ranking university (Chicago). A third of the "overhead" element of the annual grant of $80,000 to $150,000 was sufficient to fund an additional five graduate fellowships in organic chemistry and three in physical chemistry.[3] The research group at the University of Illinois was also one of the most admired. When an extended contract was requested in February 1944, it was described by some members of the Rubber Research Board as "one of the most useful and successful of the present University projects."[4]

Founded in 1867, the chemistry department at Urbana was one of the leading academic centers of chemical research in America at the time of Pearl Harbor.[5] It was preeminent in the field of organic synthesis and also strong in physical chemistry. The chairman of the department, Roger Adams, was a supporter of close collaboration between Illinois and the chemical industry.[6] A consultant for several companies, including Abbott Laboratories, Coca-Cola, and Du Pont, Adams believed that chemistry departments had to equip their postgraduate students for careers in industrial research laboratories, the number of which increased rapidly in the postwar period.

Carl Marvel had been a student of Clarence Derick (who left Illinois in 1916) and W. A. Noyes, Sr., but he worked with Roger Adams during 1918, making organic chemicals that had been hitherto imported from Germany.[7] Du Pont made him a consultant, on Adams's recommendation, in 1928. Marvel's interest in polymers was a direct result of his consulting work. Du Pont asked him to investigate the claim of the English chemist F. E. Matthews that polymers could be formed by the reaction of sulphur dioxide and ethylene.[8] Using the more convenient cyclohexene, Marvel was able to show that polysulfones were indeed formed in this reaction.[9] While it was possible to make poker chips from this material, he was unable to create a polymer sufficiently stable to be commercially useful.

This assignment spurred Marvel to study vinyl polymerization; he was particularly interested in the question of how the repeating units in the polymer were joined together. To take the case of polyvinyl chloride: were the chlorine atoms on adjacent carbon atoms (head-to-head); or on alternate carbon atoms (head-to-tail), as Hermann Staudinger had suggested;[10] or was the arrangement completely random? Marvel was able to show that polyvinyl chloride polymerizes head-to-tail.[11] By 1942, Marvel knew more about polymer synthesis than any other academic chemist, with the possible exception of Herman Mark, who had only recently come to the United States.

Much of Marvel's research for the rubber program reflected his prewar interests, for example, the mechanism of GR-S polymerization, the chemical structure of the polymer, and above all, the synthesis of hitherto unrecorded styrenes and butadienes. Even the synthesis and study of mercaptans, a key element of the wartime research at Illinois, was remarkably foreshadowed by a 1937 master's thesis by Robert Buswell (son of Illinois chemistry professor Arthur Buswell) under Marvel's supervision. Buswell investigated the reaction of mercaptans with butadiene and other dienes. He isolated an addition product, but failed to determine its structure. His poor results were partly a result of using unsuitable mercaptans, such as phenyl mercaptan and n-butyl mercaptan. Even Marvel's development of sodium polymerization can be traced back to an early interest in organometallic compounds.

The Illinois group operated in a manner different from the other research groups in the synthetic rubber program. Instead of a large group centered on the research leader, Illinois had several small groups of graduate students—largely recruited from other midwestern universities—each headed by a senior researcher. The atmosphere was informal. The senior faculty took care to know all the students and treated them with respect. The students were encouraged to do independent work but knew they could consult any member of the staff if problems arose. The various groups were loosely coordinated by Marvel, usually over morning coffee at the Farwell Cafe next to the laboratory.[12] Indeed, Herbert Laitinen and his group worked more closely with I. M. Kolthoff, who was Laitinen's former supervisor, than with Marvel. The existence of several small groups and Marvel's eagerness to obtain useful results meant that Illinois was less confined by its area of expertise than any other group in the rubber program. In the wartime period (1943–1945), Marvel's team tackled no less than fourteen discrete topics.

At the outset of the research program, Marvel attempted to find a modifier superior to Lorol mercaptan. He was especially interested in finding one that would slowly hydrolyze in the reactor, gradually but continuously liberating the active ingredient into the polymerization mixture. He hoped this would prevent the over- and undermodification of GR-S during the polymerization process. The most successful modifiers found by the Illinois group led by Robert Frank were monodecyl mercaptosuccinate and n-butyl thiolacetate, but neither was used on an industrial scale.[13] After the end of 1943—by which time the over/undermodification problem had been empirically solved by physical means and dodecyl mercaptan (a purified form of Lorol mercaptan) had shown its value—Marvel abandoned the search for a superior modifier.

Another project undertaken by Illinois in the first two years of the rubber program was the study of the inhibitory effect of possible impurities in the recycled monomers.[14] They found that 1,4 pentadiene and vinylacetylene slowed the polymerization considerably, and monoolefins had a mild retarding effect. The presence of divinylacetylene in styrene was also found to be undesirable. The Rubber Reserve Company's specifications for monomer purity were altered to reflect these findings.

In November 1942, R. R. Williams and Calvin Fuller stressed the value of studying the structure of GR-S.[15] For example, it was still unclear whether the modifier altered the amount of 1,2 addition and just how much 1,2-type polymer existed in GR-S. In the early days of the rubber program it was thought possible that many of the undesirable properties of GR-S could be connected with the vinyl side chains produced by 1,2 addition.[16]

Unsaturated compounds can be analyzed by treatment with ozone followed by decomposition of the resulting ozonides. The study of the chemical fragments produced by this "ozonolysis" can reveal the structure of the original compound. This technique was first used on natural rubber by the German chemist Carl Harries at the beginning of this century.[17] If a vinyl side chain is ozonolyzed, formic acid is produced and hence the approximate 1,2 content of a rubber can be determined by measuring the amount of formic acid present.

Norman Rabjohn used this method at Illinois in 1943. He was able to show that the 1,2 content of GR-S was about 20%, whereas sodium-polymerized polybutadiene had a higher 1,2 content, around 40%. He was also able to demonstrate that the modifier had no significant effect on the 1,2 content.[18] At Minnesota, Piet Kolthoff developed the peracid technique first used in Germany by Rudolf Pummerer in the 1920s.[19] The double bond in the main chain is attacked much more rapidly by the peracid (to form an epoxide) than the double bond in the vinyl side chain. His results agreed with Rabjohn's.[20]

Norman Rabjohn also used ozonolysis to study the arrangement of butadiene and styrene units in the polymer chain.[21] Kolthoff developed t-butyl peroxide (with a trace of osmium tetroxide) as an alternative oxidizing agent.[22] There were two major problems with the analysis of the fragments formed. Ozone is a powerful oxidant and produces "abnormal" fragments through side reactions. In any event, it was a major task to separate the fragments in a strictly quantitative manner. It is comparatively easy to use ozonolysis as a "fingerprinting" technique to show the existence of certain groups, but much more difficult to calculate the relative proportions of these groups. By extracting the acids formed and distilling their esters, Rabjohn found some evidence for the random copolymerization of butadiene and styrene.

Several years later, Marvel separated the fragments obtained from the ozonolysis of a butadiene and o-chlorostyrene copolymer using partition chromatography for the water-soluble acids and absorption chromatography for the water-insoluble acids.[23] He concluded "that styrene units are distributed according to chance and no bunching together of such units is indicated."[24]

As the production program had been established with styrene as the comonomer with butadiene, and since styrene was a cheap hydrocarbon with decided advantages as a monomer, it now appears strange that its status was ever in doubt. In 1942, however, nobody was at all sure that styrene was necessarily the best monomer. Even the fact that it was the comonomer used by the Germans was not much of a recommendation, as the German synthetic rubber was known to be of poor quality. There was a consensus that "the substitution of all or part of the styrene or the addition of a third ingredient may lead to better synthetic rubber."[25] Furthermore, this line of research was unavoidable if the production program be protected against the possibility that butadiene-styrene rubber had an irremediable but unforeseen flaw.

The synthesis of new monomers began at Goodrich and Goodyear in the middle 1930s, chiefly as a means of circumventing I.G. Farben's patents. At Goodyear, for example, James D'Ianni synthesized a large number of monomers between 1939 and 1941, including the esters of methacrylic acid and chloromaleic acid.[26] Despite Illinois's tradition of preparative organic chemistry, Marvel's group did not synthesize new monomers until the spring of 1944.[27] There were three reasons why Marvel became interested in the synthesis of monomers. He had to find a new project to replace the mercaptan work which was being wound down. Rabjohn was working on the ozonolysis of a novel copolymer prepared by D'Ianni.[28] Reilly Tar and Chemical Corporation (a major pyridine producer) had approached Goodyear in May 1943 with the suggestion that vinylpyridine be tried as a monomer. D'Ianni prepared a range of copolymers, which had some outstanding properties.[29] Unfortunately a combination of high monomer cost and problems during the curing process prevented it from becoming a useful tire rubber. It did, however, become a successful adhesive. The synthesis of new monomers at Illinois received additional impetus from Edward Meehan, at the University of Minnesota, who needed chlorostyrenes for his work on the styrene content of GR-S; the chlorine assay of the rubber provided a check on methods of styrene determination.

Marvel has given a characteristically succinct description of this research, which came to dominate the work of the Illinois group:

As good organic chemists, the Illinois group believed that they could use a new material in place of styrene with butadiene and improve the final product. Many other people in the country also felt that this was possible. But after preparing probably a hundred to hundred and fifty substitutes for styrene and trying them all, the butadiene-styrene copolymer is still used as our major all-purpose rubber. " . . . We produced probably more new synthetic rubbers during [the program] than were produced in all the rest of the government program put together . . . we did not find [a better copolymer than GR-S]—and neither has anybody else."[30]

The copolymers of butadiene and acrylic acid provide a good illustration of the difficulties faced by Marvel's group.[31] These polymers were investigated in 1952 as possible oil-resistant rubbers with better low-temperature properties than Buna N. The low-temperature behavior of the copolymers with a low acrylic acid content (5% and 10%) was acceptable, but their oil resistance was less than Buna N. Attempts to increase the proportion of acrylic acid in the copolymer were largely unsuccessful. Furthermore, while the oil resistance improved with increasing acrylic acid content, the low-temperature properties became poorer.

A year later, the Government Laboratories at Akron discovered that a copolymer of butadiene and benzalacetophenone (15%), prepared by Marvel's group, had less heat build-up than GR-S and good tensile strength.[32] As a result of this finding, the copolymers of butadiene with cinnamic acid (which is related to benzalacetophenone) and its esters were studied.[33] The cinnamic acid copolymer had outstanding tensile strength and good low-temperature properties, but had an undesirable tendency to crystallize. During the course of this research, Marvel found that the homopolymer of methyl cinnamate had good high-temperature properties, decomposing above 325°C.[34] After the program ended, he became increasingly involved in the development of heat-resistant polymers, which culminated in the commercialization of PBI heat-resistant fiber in 1983.

The grounds of styrene's success were largely economic: it was simply the cheapest monomer that was suitable. Herman Mark and Carl Wulff, at I.G. Farben's Ludwigshafen plant, had developed an industrial process for the manufacture of styrene from coal tar in 1929.[35] The Udex process developed by Universal Oil Products and Dow Chemicals made cheap petroleum-based benzene (and hence styrene) readily available in the mid-1950s.[36] Furthermore, the concurrent consumption of styrene in polystyrene manufacture permitted the construction of large plants with consequent economies of scale. Alternative monomers might provide a technical improvement in performance over GR-S, but this would be counterbalanced by their higher cost.

Chicago: Free Radicals and Emulsions

The elucidation of the course of the polymerization of styrene and butadiene was the most important of the various tasks allotted to the universities. The solution of this intricate problem (which was only partly solved by the end of the program) involved the study of the emulsion and the chemistry of the free radicals formed during the polymerization. The University of Chicago had two teams working on this question, one headed by Morris Kharasch, which studied free radicals, and the other led by William Draper Harkins, which investigated the role played by the emulsion.

Harkins joined the Chicago faculty in 1912, and he became professor of physical chemistry five years later.[37] From the very beginning, he was interested in atomic structure and nuclear reactions. For many years Harkins was one of the few American chemists who pursued nuclear research. Harkins also had a long-standing interest in surface chemistry, which stemmed from his research with Fritz Haber in 1909 and with A. A. Noyes, then still at the Massachusetts Institute of Technology, a year later. Harkins carried out a series of very accurate surface and interfacial tension measurements between 1917 and 1930. This research was carried out in strenuous, even frantic, competition with Irving Langmuir, who was awarded the 1932 Nobel Prize that Harkins believed was rightly his. Harkins retired from teaching in 1938 at age sixty-five but continued his research until his death in 1951.

After taking his bachelor of science degree at Chicago in 1917, Morris Kharasch had worked on war gases in the Gas-Flame Division of the U.S. Army, which stimulated his interest in organometallic chemistry.[38] The chemistry department at the University of Chicago had a tradition of physical organic chemistry, stemming from Julius Stieglitz's research on organic rearrangements in the early years of this century.[39] Kharasch worked on the preparation of organomercury compounds for his doctoral research in 1919. He later used his doctoral research to assist companies with the development of drugs and pesticides. After his graduation, however, Kharasch was more interested in using the acid scission of organomercury compounds with two different groups to determine the electronegativity of organic radicals. In his research on organomercurials, Kharasch was assisted (at different times) by Frederick Stavely,[40] who became Firestone's research director, and Russell Marker,[41] who later laid the foundations of the modern steroids industry.

As a result of his work on the electronegativity of organic groups, Kharasch became interested in the anomalous (and exceedingly erratic) addition of hydrogen bromide to allyl bromide. With his student Frank Mayo, Kharasch published a paper in 1933, which demonstrated that the variation in the products of the addition was due to traces of peroxides present in aged samples.[42] Four years later, after much painstaking research, Kharasch announced that the anomalous addition was a free radical reaction.[43] In the same year, a young chemist at Du Pont, Paul Flory, published a four-stage scheme for free radical polymerization, despite having no background in free radical chemistry and without any significant literature to help him.[44] This basic scheme still stands today.

Displaying a remarkable (if serendipitous) prescience, Kharasch proceeded to carry out research that was to be of considerable value to the rubber program, in addition to his extensive work on chlorination/bromination and Grignard reagents. Kharasch's collaborators at this time included—in addition to Frank Mayo—Frank Westheimer, Herbert C. Brown, and Cheves Walling. In 1938, he showed that the addition of mercaptans to olefins, which had baffled young Buswell, was a free radical reaction.[45] Kharasch studied the effect of oxygen on free radical reactions, and by 1943 he was convinced that even traces of oxygen could inhibit free radical polymerization.[46] In 1941, he began to study the catalytic decomposition of organic peroxides. This research, which was mainly carried out by Walter Nudenberg, came into its own with the introduction of cumene hydroperoxide in the cold rubber recipes.

In 1942, it was not clear how the mercaptan modifiers influenced the polymerization of GR-S. The concept of "chain transfer" had been introduced by Paul Flory in 1937, but it had not been applied systematically to a significant body of experimental data. In chain transfer, a suitable molecule donates an atom to a growing polymer chain, which halts the polymerization and converts the donor molecule into a free radical. The new free radical then attacks a monomer molecule, initiating a new polymer chain. This produces shorter chains, and hence lowers the average molecular weight of the polymer.

Kharasch soon realized that the mercaptan, with a weak sulfur-hydrogen bond, was acting as a chain transfer agent.[47] Robert Carlin at the University of Illinois independently arrived at the same conclusion on the basis of Charles Price's research on the emulsion polymerization of styrene. Price found that the initiator and the chain transfer agents became the end-groups of the polystyrene.[48] Carlin discovered that this was also true in the case of GR-S and that the slow step in the reaction was the initiation step.[49] Both groups showed that excess modifier produced low molecular weight polymers and ultimately a simple addition compound was formed when a 1:1 molar ratio was used.

The General Laboratories of U.S. Rubber on the top floor of a converted silk mill in Passaic, New Jersey, played a significant role in this research. Frank Mayo, who had left Chicago to join U.S. Rubber in 1942, was the driving force behind this program of research on free radical polymerization, which also reflected U.S. Rubber's traditional bias toward physical chemistry.[50] The company enjoyed a close relationship with the University of Chicago. W. D. Harkins was a consultant for the company before the rubber program was set up. Mayo maintained close contact with his former colleague Morris Kharasch, and another Kharasch student, Cheves Walling, joined Mayo's group in 1943. Ironically, this advantage in fundamental research did not translate into a competitive edge in innovation.

Mayo published an important paper on chain transfer in 1943.[51] He was aware of the research on modifiers by Carlin and Kharasch, and its implication for chain transfer generally. Unable to publish work on modifiers because the rubber research was confidential during the war, Mayo turned to the study of chain transfer by the solvent. Certain solvents can act as chain transfer agents, in the same way as mercaptans. Hence, the polymerization of styrene in such solvents should reduce the final molecular weight, and the change in molecular weight will vary from solvent to solvent. Mayo found both of these predictions to be correct. On the basis of Flory's theory and his own experimental work,

Mayo introduced the concept of a transfer constant, which was related to the ability of a given solvent (or modifier) to terminate a growing chain. The transfer constant was not only a good way of comparing the activity of different solvents (or modifiers), but it was also much easier to determine experimentally than the actual rate constants.

After the early ground-clearing work by Kharasch and Carlin, the focus of the research shifted to the study of the reaction site. Did the reaction take place in the aqueous phase, the hydrocarbon phase, or at the interface? To answer this question, it was necessary to understand the structure of the emulsion. Harkins used dyes to measure the critical concentration for micelle formation, and small-angle X-ray diffraction to determine the diameters of the micelles. By 1946, Harkins was able to work out the structure of the emulsion and its relationship to the course of the reaction.[52] The initial free radicals are found in the aqueous phase, and as Kharasch had discovered earlier, they are relatively stable. Harkins suggested that they migrate to soap micelles which contain small droplets of monomer. At about 13% conversion, these particles are exhausted and the polymerization continues in polymer particles swollen with monomer. The large monomer drops dispersed throughout the emulsion take no significant part in the reaction but act as reservoirs that supply the reaction sites with fresh monomer.

By the war's end, Kharasch had shown that the course of the polymerization of butadiene and styrene was controlled by the reactivity of the relatively stable free radicals formed by the addition of the initial free radicals to the monomers, rather than the monomers themselves. Kharasch and his group then studied the stabilization of the monomers, especially the spontaneous formation of popcorn polymer by recycled butadiene.[53] In 1947, Nudenberg and Kharasch polymerized butadiene with an ordinary Mazda lamp (using uranyl acetate as a photosensitizer), but the product was no different from polymers produced by more conventional means.[54]

During the postwar period, Harkins's model of emulsion polymerization was used to derive a more exact analysis of the mechanism and kinetics of emulsion polymerization. In 1947, Harkins's coworker Myron Corrin produced a treatment of the kinetics of styrene polymerization.[55] The following year a more complete analysis, divided into various special cases, was published by Wendell V. Smith and Roswell Ewart of U.S. Rubber, who were working in close collaboration with the two Chicago groups.[56] Like Corrin, they considered only the polymerization of styrene, but Maurice Morton was able to show in 1952 that the emulsion polymerization of butadiene fitted one of the special cases analyzed by Smith and Ewart, if it was carried out at low temperatures.[57]

Between 1950 and 1956, Morris Kharasch and Walter Nudenberg

investigated the acid-catalyzed and thermal decomposition of hydro-peroxides, which were used in the cold rubber recipes. In 1950, they discovered that cumene hydroperoxide could be converted into the more stable cumyl peroxide.[58] In their subsequent work, they patiently elucidated many of the complexities surrounding the reaction between hydroperoxides and iron or cobalt salts.[59]

Minnesota: Analysis and Kinetics

The methyl rubber produced by the Germans during World War I had a polymerization time of several months. The mutual recipe took twelve hours at 50°C, but several days at the more desirable temperature of 5°C. A rapid polymerization process was fundamental for the development of cold rubber. To make the polymerization faster, it was necessary to measure the effect of changing the parameters of the mutual recipe. The rate of any reaction can be calculated by measuring the concentration of the reactants in the mixture at set intervals. Chemical analysis therefore makes an important contribution to kinetics, the study of the rate of chemical reactions. In the case of GR-S, the rate of polymerization was slow enough to make the sampling process straightforward, but the concentration of several chemicals, especially the initiator and the modifier, was very low. Izaak M. (Piet) Kolthoff at the University of Minnesota in Minneapolis, a leading analytical chemist, was invited by R. R. Williams to find the best methods for their determination.

While training as a pharmacist at the University of Utrecht in the Netherlands, Kolthoff worked with a highly original professor of analytical chemistry, Nicholaas Schoorl, and took his doctorate in this subject in 1918.[60] Like Schoorl, he was keen to apply the new insights of physical chemistry to the more staid field of chemical analysis. His initial interest was in acid-base titrimetry and similar volumetric methods of analysis, but he also studied electrochemical methods. Kolthoff published a book on conductometric titration in 1927 and a two-volume monograph on volumetric titration in 1927–1928.

Hugo R. Kruyt, a noted colloid chemist at the University of Utrecht, visited America in the spring of 1927, and was guest of honor at the fifth national symposium on colloid chemistry at the Massachusetts Institute of Technology. During his visit, Kruyt was asked by Samuel Lind of the University of Minnesota to suggest a good chemist to assist with the graduate research program. He recommended his close colleague Kolthoff, who joined the chemical faculty at the University of Minnesota a few months later.[61]

In 1932, Kolthoff resumed his study of precipitates, a topic he had initially investigated at Utrecht in the early 1920s. Precipitates, insoluble salts deposited from solution, are important in chemical analysis as a means of estimating the concentration of metals. However, they often contain impurities, usually as a result of coprecipitation, and they change their characteristics with time ("aging"). Lind suggested the use of natural radioactive lead ("thorium B" from radium) as a means of studying the aging of lead sulfate. This was one of the earliest examples of radioactive tracers in chemical analysis.

Jaroslav Heyrovsky, from Charles University in Prague, Czechoslovakia, visited Minnesota in 1933 and aroused Kolthoff's interest in polarography.[62] This technique, which Heyrovsky had pioneered in the mid-1920s, measured very low concentrations of a wide variety of materials by tracking their reduction at a mercury electrode. Kolthoff and his student James Lingane transformed this esoteric technique into a conventional analytical method. Their monograph, *Polarography* (New York: Interscience, 1940; second edition, 1952), did much to popularize it in the United States.

In 1939, Lingane worked briefly on an electrochemical (amperometric) titration of sulfur-containing proteins, with the aim of developing a blood test for cancer.[63] The amino-acids that contribute the sulfur to proteins, such as cysteine, are chemically related to mercaptan modifiers. Kolthoff realized that this technique could be used to measure the low mercaptan concentrations in the mutual recipe.[64] It was a cheap and simple method, which used only apparatus available in many laboratories. He also developed a polarographic method for the measurement of persulfate concentrations.[65] Of course, Kolthoff employed nonelectrochemical techniques, such as volumetric titration, when they produced reliable results.

Having established a sound analytical basis, Kolthoff and his group studied the kinetics of the mutual recipe. With Wesley Dale, Kolthoff showed that the rate of polymerization of styrene with potassium persulfate, in emulsion, was often independent of the styrene concentration but proportional to the square root of the persulfate concentration.[66] This kinetic equation, which can be derived from a simple (steady-state) model of free radical polymerization, was independently confirmed by Charles Price and Clark Adams at the University of Illinois.[67]

Kolthoff and Walter Harris investigated the behavior of the modifier.[68] They demonstrated that the rate of disappearance of the primary dodecyl mercaptan and the tertiary dodecyl mercaptan was similar, whereas primary octyl mercaptan was consumed much more rapidly than tertiary octyl mercaptan. The two dodecyl mercaptans were the best modifiers in the primary and tertiary series. However, primary dodecyl mercaptan (DDM) was much more sensitive to outside influences, such as agitation. Kolthoff thus confirmed the value of DDM, but also indicated the superiority of its tertiary analogue.

Soap is produced by the reaction of organic "fatty" acids with sodium hydroxide. As mentioned in the previous chapter, certain fatty acids created production problems in the first year of the program. Kolthoff and his group examined the effect of excess soap, sodium hydroxide, and fatty acids, in turn. Below a threshold level, the initial rate of polymerization was roughly proportional to the soap concentration, but the eventual maximum rate was completely independent.[69] Excess sodium hydroxide had no effect on the rate, but the addition of excess fatty acid accelerated the rate of polymerization. This indicated that the soap and the fatty acid were equivalent as far as the rate of polymerization was concerned. In fact, the rate remained unchanged when 20% of the soap was replaced by free acid.[70]

Whereas all organic compounds have bands in the infra-red region of the spectrum (see Intermezzo in Chapter 2), only certain classes of compounds have strong bands in the ultra-violet. The GR-S units derived from styrene have a band associated with the benzene ring. Using the then novel DU Beckman spectrophotometer, made by National Technical Laboratories, Kolthoff's colleague Edward Meehan observed how the styrene content varied as the polymerization progressed to completion.[71] He found that the composition at a given conversion of monomers to polymer fitted a simple curve. This started at a low styrene content, related but not equal to the initial ratio of styrene to butadiene (the charge ratio). The overall average styrene content increased until it was equal to the charge ratio at the end. The composition was independent of temperature and the other components of the recipe. Meehan's work indicated one reason why it was undesirable to allow the polymerization to proceed to completion. The final segment of the completed polymer chain was an inelastic resin with a high styrene content that lowered the overall quality of the rubber.

In early 1944, William Reynolds at the University of Cincinnati found that a class of compounds called diazo-thioethers accelerated the emulsion polymerization of butadiene and styrene with potassium ferricyanide.[72] Kolthoff and Dale investigated this reaction and discovered that p-methoxyphenyl, diazothio-(2-naphthyl)-ether (MDN) was by far the most effective catalyst in this group. Unlike potassium persulfate, MDN could be used with rosin soaps, and the alkaline recipe developed by Kolthoff and Dale was the first formula to give useful results at 5°C.

In 1946, after the German work had been published, Minnesota and Hercules independently developed the iron-pyrophosphate-cumene hydroperoxide-sugar-rosin soap recipe discussed in the last chapter. If the iron is in a freely soluble form (usually ferrous sulfate), there is a

rapid reaction between the iron and the peroxide and hardly any polymerization takes place. However, if the active concentration of iron
is low, the reaction with the peroxide is moderated and polymerization
takes place. Sodium pyrophosphate, which forms a barely soluble compound with ferrous sulfate, was an ingredient of the recipe offered to
the rubber program in 1941. This salt was used with dextrose sugar in
the standard Custom recipe and in the sugar-free recipes developed by
Phillips in 1949–1951.[73]

The original sugar-free recipes worked best at –10°C, but this required the addition of methanol as an anti-freeze. Sugar-free recipes that
used rosin soaps and potassium pyrophosphate-ferrous sulphate—and
operated at 5°C—displaced the Custom recipe during the mid-1950s.
Cumene hydroperoxide was gradually replaced by the more reactive di-
isopropylbenzene hydroperoxide and p-menthane hydroperoxide.

Piet Kolthoff and Avrom Medalia tried to replace the sugar, which
sometimes gave erratic results, with other reducing compounds. When
they used sodium sulfide, the formation of ferrous sulfide turned the
solution deep blue.[74] The insoluble sulfide had the same effect as the
pyrophosphate complex. U.S. Rubber used this recipe to make high-
solids GR-S latex for paints in June 1949.

This method of making latex was displaced, however, in the mid-
1950s by the "peroxamine" recipes first developed in 1950 by George
Stafford Whitby at the University of Akron.[75] These recipes contained
a peroxide (such as cumene hydroperoxide) and a polyamine, for example, tetraethylene-pentamine (TEPA). The active species was actually a complex of TEPA and iron present as an impurity. Since the
amount of iron involved was minute, the recipe was practically "iron-
free." This was useful, as iron can accelerate the aging of rubber.[76]

A complex between iron and a strongly binding compound (che-
lating agent), such as ethylenediamine-tetraacetic acid (EDTA or Ver-
sene), has the same effect as an insoluble iron compound: the free iron
concentration is very low. EDTA was first used by the Polymer Corporation at Sarnia in Ontario in 1948.[77] It developed the "sulfoxylate"
recipe, which replaced the pyrophosphate of the Custom recipe by
EDTA, dextrose by formaldehyde, and cumene hydroperoxide by
p-menthane hydroperoxide. Kolthoff and Meehan used EDTA in their
"Veroxasulfide" recipe, which contained di-isopropylbenzene hydroperoxide, sodium sulfide as the reducing agent, and the complex of ferric
nitrate and EDTA.[78] The iron content was a hundred times less than
the level used in the Custom recipe.

Clearly, it was important for the development of cold rubber to
elucidate the reaction between iron and peroxide. In 1948, Kolthoff and
Medalia began a detailed study of this reaction.[79] In the course of this

research, Kolthoff became interested in the effect of adding ethanol or acetic acid. No reaction occurs with hydrogen peroxide and ethanol alone, but ethanol is rapidly oxidized by a mixture of ferrous sulfate and hydrogen peroxide (Fenton's reagent).[80] This is called an "induced" reaction: the oxidation of ethanol by peroxide is induced by the iron. However, this reaction is suppressed if acetic acid is added to the mixture.[81] Similarly, the extent of reaction between cumene hydroperoxide and the ferrous ion on their own is much less than theory would indicate. Kolthoff and Medalia found that the reaction was promoted by the addition of ethanol. They were able to use this phenomenon to develop an accurate method of measuring concentrations of organic peroxides.[82] This is a good example of the continual interaction between chemical analysis and kinetics.

Cornell: Physical Methods and Polymer Configurations

The two Nobel laureates at Cornell University (Debye, 1936; Flory, 1974) shared a common methodology of rigorous theory and skilled experimental technique. They were both interested in the size and shape of polymer molecules. They both took earlier work and developed it into a set of new theories and techniques. Nevertheless, there were differences between the two men. Polymer science was incidental to Debye's main interests, a coda to a fruitful life spent in chemical physics, but it was Flory's vocation. Consequently, Debye made a major contribution to one field, light scattering, while Flory made equally important contributions to many—almost all—areas of polymer science. Whereas Debye worked within the framework of the rubber research program, Flory transcended it.

The celebrated colloid chemist Wilder Bancroft had molded the chemistry department at Cornell, hitherto very much an agricultural college, into a center for his individualistic version of physical chemistry in the early part of this century.[83] However, the chemistry department fell into disarray after his almost fatal accident in 1937 and the formation of a separate chemical engineering department in the same year. It was rescued by the appointment of John Kirkwood as Todd Professor of Chemistry in 1938.[84] Kirkwood had worked with Peter Debye in 1931. When Debye arrived in the United States in early 1940, after fleeing Germany, Kirkwood invited the Dutchman to be the George Fisher Baker Lecturer. This is a temporary lectureship which is awarded every year to permit an eminent chemist to deliver a series of lectures at Cornell. Debye was then appointed head of the department, a position he retained until 1952.[85] He stayed at Ithaca until his death in 1966.

Debye was the greatest living expert on the electrical properties of the molecule. Bell Telephone Laboratories (BTL) had sought to hire him as a consultant to help with the development of high-frequency insulation, but the Laboratories' president was warned that Debye would be refused security clearance because he had relatives in occupied Europe.[86] Debye worked closely with the BTL group during the research program, but during the war he had to be escorted by uniformed policeman whenever he entered a defense plant.[87] He also had a long-standing interest in the interaction between electromagnetic radiation and matter. This resulted in the development of the X-ray diffraction of powdered solids and gases, which revolutionized our understanding of chemical structure.

When the rubber program began, new methods for the determination of the molecular weight of very large polymers, especially the undesirable cross-linked gel molecules, were sorely needed. Contemporary techniques, such as viscosity and osmotic pressure measurements, were insensitive at high molecular weights. The viscosity-derived molecular weights were also known in general to be too low.

The measurement of the light scattered by a polymer solution offered a way of achieving this objective if the necessary equations could be derived. The basic theory had been developed by the 3rd Baron Rayleigh to explain the blue color of the sky in 1871.[88] This equation was not directly applicable, however, to solutions. In the mid-1930s, Samuel Gehman and John Field at Goodyear had attempted to use the related Raman scattering to study natural rubber, but found it was too weak. They then turned to light scattering and measured the size of the rubber molecule in 1937.[89] Unfortunately, they used Staudinger's theory of rigid polymer molecules and estimated an excessively low molecular weight of 53,000 from the observed chain length of 3,600 angstroms.

In March 1943, Debye realized that a paper published by Albert Einstein in 1910[90] provided a solution to the problems encountered by Gehman and Field. Einstein had assumed that light scattering in solutions was produced by fluctuations in the concentration of the dissolved substance (solute). This fluctuation is related to the osmotic pressure and hence the molecular concentration of the solute. Once the relationship between the concentration and the refractive index of the solution was known—which is easy to measure—the molecular weight of the solute could be derived from the measurement of the solution's turbidity at different concentrations. Debye worked on this problem with the assistance of William O. Baker's group at Bell Telephone Laboratories. He laid out the fundamental equations in a report to the rubber research program in July 1943, which was later published in the *Journal of Applied Physics*.[91] He then refined this initial result by taking various sources of error into account, such as depolarization.[92]

Debye's small group used these equations to calculate the weight and size of various polymers. For example, his student Fred Billmeyer discovered in 1944 that the size of a polyvinyl chloride molecule in solution lay between the extremes of a random coil with free bond rotation and a rigid rod.[93] In collaboration with Roswell Ewart and Charles Roe of U.S. Rubber, Debye investigated the possible use of turbid solutions produced by adding a precipitant to the original solution (for example, methanol to benzene).[94] However, they found that difficulties arose if the solvent and the precipitant had markedly different refractive indices.

In the late 1940s and early 1950s, Debye used the light scattering technique to study solid polymers (polymethyl methacrylate),[95] polar polymers (poly (p-chlorostyrene)),[96] basic polymers (polyvinylpyridine) in acidic solvents,[97] and soap micelles.[98] He was particularly interested in the determination of the shape of polymer molecules and micelles. His two leading coworkers in this period were brothers: Arthur (Art) and Frederick (Fritz) Bueche. Debye and Fritz Bueche studied the diffusion of individual polymer molecules through the solid polymer. They were able to show in 1952 that the product of the self-diffusion constant and the bulk viscosity was a constant, in accordance with theory.[99]

Debye and Arthur Bueche had studied the viscosity, thermal diffusion, and sedimentation of polymer solutions in the mid-1940s.[100] These phenomena all display the same basic relationship to the molecular weight of the polymer, for example:

$$[\eta] = KM^{a}$$

where $[\eta]$ is the intrinsic viscosity, K is a constant for that solution, and M is the weight average molecular weight.[101] On the basis of his theory of rigid rod-like polymers, Staudinger had predicted that $a = 1$.[102] Experiments had shown, however, that a usually lay between 0.5 (the square root) and 1. Debye developed a model that approximated the polymer coil (first proposed by Werner Kuhn in 1934[103]) to a sphere containing a variable number of rigid balls.[104] This led to the theoretical result that a should be 0.5. This error arose because of Debye's failure—in common with almost all other polymer scientists—to allow for the excluded volume effect, which exists because a real polymer molecule, unlike a theoretical coil, cannot overlap itself.

This error was remedied by Paul Flory, who had just become the Baker lecturer. By taking the excluded volume effect into account, he was able to show that the constant involved (K in the above equation) was the same for most polymers and solvents.[105] Furthermore, the "ideal" case of $a = 0.5$ would occur in certain poor solvents, which he called theta (Θ) solvents. He also demonstrated the importance of the critical miscibility temperature for polymer solutions, similar to the Boyle temperature for gases.[106] He named this the theta point, but it is now called the Flory temperature. His theoretical advances also permitted Flory to calculate the molecular dimensions of a polymer in a theta solution.[107] These measurements could be compared with the theoretical length, based on free rotation, or the dimensions obtained from light scattering. Like Billmeyer, he found that the actual length was usually much greater than the theoretical prediction.

Flory's nine years at Cornell were a very fruitful period for the intellectual development of polymer science, similar to Thomas Hunt Morgan's tenure at Columbia University for genetics. With his colleagues and students, including Thomas Fox, Harold Scheraga, William Krigbaum, and Leo Mandelkern, Flory reexamined viscosity,[108] sedimentation,[109] and diffusion[110] on the basis of his new equations for polymer solutions. He also found the time to revise his Baker lectures and publish them as *Principles of Polymer Chemistry* (1953), which quickly established itself as a scientific classic. Flory's Nobel Prize in 1974—only the second for advances in polymer chemistry—was awarded for his pathbreaking research at Cornell.

The Polymer Research Discussion Group

As the usual channels of communication, above all the publication of results in scientific journals, were prohibited during the war on the ground of national security, the academic chemists required a means of exchanging information. Furthermore, there was a need to bring the academic and industrial groups into a close relationship. Even before the wartime research program began, R. R. Williams and Calvin Fuller had noted the geographical separation of the research groups and proposed the formation of "a discussion group whose function will be mutual stimulation and exchange of ideas." They went on to say that:

The meetings of the research discussion group should be in the nature of an informal, unbiased consideration of research results lead by one of the members close to the subject under discussion. Definite agenda should be prepared, and followed and in so far as possible all members should be fully informed before hand on the subjects to be considered.[111]

This is essentially how the discussion group did function when it was formed in December 1942. It brought the leading scientists together for a two-day meeting every three months during the war and twice a year thereafter. The meetings were commonly held in Akron or New York, and occasionally in Chicago, Washington, or other cities. Papers were presented and discussed at the meeting, and there was also a more general discussion of the problems facing the research program.

The first meeting of the discussion group (it was initially called the Rubber Research Discussion Group, but the name was soon changed to the Polymer Research Discussion Group) was held in Akron on 28 and 29 December 1942.[112] In May 1943, the Polymer Research Policy Committee fixed the membership of the group as four representatives "from each Company laboratory and three from each University laboratory."[113]

The discussion group was not a wholly novel idea. It took over several of the functions of the Rubber Reserve Company's Technical Committee formed in 1940. This consisted of the representatives of the participating companies, and was responsible for the assessment of the technical information given to the RRC. In the research program, its function as a policy-making body was initially taken over by the Policy Committee, which was disbanded at the end of 1943. The discussion group replaced the Technical Committee as the forum for the discussion of research results.

The discussion group also took over the role played in the prewar period by the symposia organized by the American Chemical Society's (ACS) Rubber Division. The Rubber Division was more autonomous than the other ACS divisions, and its meetings had a strong bias toward industrial research. The Rubber Division also sponsored publications, including two that bracket the program: *Chemistry and Technology of Rubber* (1937), edited by C. C. Davis and John T. Blake, and *Synthetic Rubber* (1954), edited by G. S. Whitby, C. C. Davis, and R. F. Dunbrook. In 1940, the Rubber Division had sponsored a symposium on "Rubber for Defense" at the ACS spring meeting in Atlantic City.

If rubber chemists are still cut off to some extent from other chemists, it was even more the case in the days when rubber chemistry was still one of the "black arts," and the division did much to maintain a strong esprit de corps. James D'Ianni has remarked:

The early days of the rubber industry were marked by much secrecy with respect to exchange of technical information. Gradually, management realized that more was to be gained than lost by allowing its technical experts to discuss problems with their peers in other companies and laboratories. The Rubber Division has been the single most important catalyst in effecting such discussions without loss of confidentiality, as well as promoting camaraderie and life-long friendships.[114]

Perhaps the closest precursor of the discussion group was the conference held every summer on Gibson Island, Maryland, until 1947. The venue was then transferred to the hills of New Hampshire, because of the humidity and heat of Gibson Island. The accommodations there were also too small for the larger postwar conferences.[115] The conferences were also renamed the Gordon Research Conferences, after the founder of the *Journal of Chemical Education,* Neil Gordon, who had organized the first conference in 1931.[116] At these meetings, the best scientists in a particular field met under informal conditions to discuss the latest results and to propose possible lines of future research. The papers were restricted to genuinely novel material: reviews were not permitted. Comments from the audience (and postmeeting discussions in the bar or on the beach) were of great value. Frankness of discussion was promoted by a prohibition on publication of the conference's proceedings. The early conferences were relatively small, with less than a hundred scientists being invited to each conference.[117]

The discussion group was also a forum for the candid exchange of results. It, too, helped to break down the barriers between companies and laboratories; above all, it was a bridge between academe and industry. During the war, it created a strong spirit of friendship and cooperation between different and hitherto unconnected groups of scientists. Whereas the other meetings we have discussed assumed expert knowledge, the discussion group also had a didactic function. The academics brought the industrial researchers up to date with the latest developments in polymer science and physical chemistry, and the rubber technologists explained some of the mysteries of their craft.

The Universities and Industrial Innovation

In the academic research we can observe the same distinction between incremental innovation and radical innovation that we have noted in the industrial research. The university-based groups made significant contributions to the development of GR-S, for example free-radical mechanisms, models of emulsion polymerization, structure of GR-S, and analytical techniques. They were less successful with the innovation of commercially viable synthetic rubbers. We have noted the failure of Avery Morton's Alfin rubber, Marvel's sodium rubber, and Kharasch's lithium rubber. It could be argued that universities do not provide an appropriate environment for radical innovation. Of course, competition exists between academic groups, especially for research funds, but the yardstick is publications in scholarly journals rather than profits. Most academic scientists seek originality and peer esteem, not new products and technical advantage. Given the complex nature of scientific research, this is obviously an oversimplification, and there is clear evidence that universities can create radical innovations. In the field of polymers alone, Karl Ziegler and Giulio Natta developed linear polyethylene and polypropylene, Carl Marvel initiated a wholly new class of heat-resistant fibers (PBI), and Charles Price created the first polyurethane based on propylene oxide.[118]

Melvin Kranzberg has suggested that mission-oriented research, which responds to specific technical needs, tends to produce incremental innovations. By contrast, technological breakthroughs often stem from non-mission-oriented scientific discoveries.[119] As examples of "scientific-pull," Kranzberg mentions airflight, nuclear energy, and rocketry. They were all innovated by civilian scientists or engineers before their military applications became evident. Kranzberg remarks, "No major—and I stress the word 'major'—scientific discovery or technological advance seems to have been derived from the push of military requirements."[120] These military requirements led, however, to the *improvement* of aircraft, missiles, and electronic devices.

Kranzberg's generalizations can be applied to the case of synthetic rubber. Buna S/GR-S was initially developed independently of the need for a domestic supply of rubber in wartime, but military requirements led to its improvement in the form of cold rubber and low-temperature-resistant varieties. Similarly, the American synthetic rubber research program failed to anticipate the breakthroughs made by Karl Ziegler, who was working in an institute nominally dedicated to coal research. Kranzberg's argument is thought-provoking, and its application to the development of synthetic rubber could be extended.

Recent studies of technological innovation have often stressed the importance of the relationship between different communities, for example, between scientists, industrialists, and government officials.[121] This approach suggests that the root cause of the universities' limited success may be found in the interaction between the academic research groups and the rubber companies. There was a general perception, within and outside the rubber industry, that the universities were being invited to join the program only to assist with the development of the scientific underpinning of the industrial processes.[122] Many of the academic groups operated in harmony with this division of labor. Indeed, some groups—Debye's and Harkins's come especially to mind—would have found it difficult to essay technological innovation.

Obviously, most of the academic groups initially knew very little about the manufacture and properties of rubber, and even less about its consumers. Studies of innovation have shown that a lack of understanding of the market for an innovation is very often fatal to its chances of success.[123] While their understanding of rubber manufacture and the significance of the final product's physical properties increased considerably in the course of the program, the academic groups were never close to the consumers.

This weakness was accentuated by the rubber industry's cool response to the universities' early enthusiasm. In contrast to the major chemical and pharmaceutical corporations, the rubber companies had hitherto lacked a close relationship with academic scientists. Kolthoff has described the position at the beginning of the program:

Our whole problem was that we simply did not know what the problems were. No, I am not joking. In the very beginning, I said to Maurice Visscher, "I'm so damned mad. I'd like to publish that the big companies, Goodyear, Goodrich, Firestone, all disliked the idea that university people would come in and stick their noses in their business." We didn't know what the problems were. You asked what problem? We didn't have a problem. We didn't know the problem. Surely, we could analyze those things and so I recall that [Edward Meehan] and I went together to Goodyear and Goodrich and were told, "We want to know the purity of the standard substances which are being used to make rubber, but there is a committee working on them already so we don't need you for that."[124]

The trepidation with which some research leaders in the rubber industry greeted the entry of the university scientists is reflected in an internal company report for August 1943, written by the head of Goodrich's synthetic rubber research section, Harlan Trumbull:

This field of modifiers is a Roman holiday for the academic scientists and they are crowding to the circus, but strangely they do not know rubber very well and it is becoming a game of hide the pill. Out of it all is destined to come encyclopedic publication or Edisonian invention on the part of those who "darken counsel by using words without knowledge."[125]

What were the reasons for the industry's lack of enthusiasm for cooperative research with the academic groups? The rubber experts were hard-pressed and were naturally unwilling to take the time to explain their concerns to academic chemists. Furthermore, the rubber industry had a strong sense of self-sufficiency. This came across very clearly when I interviewed chemists who worked in the rubber industry during the program. The rubber industry personnel saw no compelling need at the time for the academic researchers, although the contributions that were made by I. M. Kolthoff, W. D. Harkins, and William Reynolds were generously acknowledged. This initial disdain diminished considerably over the lifetime of the program, but the rubber companies nevertheless felt no compelling need to assist the academic researchers or to pursue the leads they created. This was serious for the universities, for they were hardly in a position to do the necessary scaling up and troubleshooting on their own. The research section of the Office of Rubber Reserve was not anxious to make a delicate situation worse by insisting that the rubber companies work more closely with the universities. Avery Morton was perhaps an exception, because Goodyear was interested in his Alfin catalysts.[126]

The academic research program did not lead to long-term collaboration between the universities and the rubber companies. It is striking that none of the major universities continued to collaborate with the rubber companies after the rubber program. Nor did it lead to any significant number of academics or Ph.D.'s entering the rubber industry.[127] It is difficult to disaggregate the influence of the academic research on the industrial research from the concurrent effects of synthetic rubber manufacture and the burgeoning of polymer science. It is clear, however, that the rubber companies did undertake more fundamental research after 1942. Kharasch and Harkins had a particularly strong influence on the basic research program at U.S. Rubber. Nevertheless, Du Pont's remarkable commercial success following its 1927 decision to pursue fundamental research probably provided greater encouragement.

At the beginning of the program, Piet Kolthoff was an unfamiliar figure to the rubber companies and we have seen how he found it difficult to find out what the problems were. Yet he became the most successful of the academic scientists in the program and perhaps the most respected. He is still held in high regard by the men who worked in the industrial laboratories. Kolthoff's reputation surpasses even Marvel's, whose team received the most funds of any academic group. What were the reasons for Kolthoff's success?

As an immigrant, Kolthoff was accustomed to easing himself into new situations. He patiently built up respect for his research by developing new analytical techniques and experiments that helped to solve the industry's problems. Having won the industry's gratitude and respect, he moved into new areas, such as the hydroperoxides and rosin soaps. Part of Kolthoff's success was the nonthreatening nature of his research. While he moved into new areas, he did not seek to dominate them in a way that worried the rubber companies. Nor did he attempt to create wholly new polymerization techniques like Marvel and Avery Morton. Each step was a logical extension of his earlier work. Indeed, they were often topics taken up by the rubber companies, who then ran into difficulties. Kolthoff helped to solve these problems or at least elucidate them by methodical experimental research. Kolthoff's keen sense of the industry's needs and his willingness to enter new research areas were key elements of his success.

The university-based research during the war was one of several reasons why the American GR-S was better than the German Buna S in 1945.[128] Within the narrow confines provided for it by the rubber industry, the academic research program was successful. However, the cooperative research did not stimulate technological breakthroughs by the universities or result in new synthetic rubbers. The collaboration between the universities did not materially accelerate the development of cold rubber or synthetic "natural rubber." Furthermore, there is little evidence that the university-based research significantly strengthened the technological position of the American synthetic rubber industry in the late 1950s and 1960s.

Notes

1. Calculated from table in Solo, *Synthetic Rubber,* 97.

2. Arnold Thackray, "University-Industry Collaboration and Chemical Research: A Historical Perspective," in *University-Industry Research Relationships* (Washington, D.C., 1982), 193–223. Jeffrey L. Sturchio, "Chemists and Industry in Modern America: Studies in the Historical Application of Science Indicators" (University of Pennsylvania Ph.D. thesis, 1981), 205–212.

3. "The University's Synthetic Rubber Program During World War II," November 1956, UI 39/1/8 Box 1, 19.

4. "Summary of . . . Meeting No. 14," 9 February 1944, RG 234, RRC, Entry 231, PI–173, Minutes of the Rubber Research Board, 4.

5. P. Thomas Carroll, "Perspectives on Academic Chemistry in America, 1876–1976: Diversification, Growth and Change" (University of Pennsylvania Ph.D. thesis, 1982), especially Chapter Four, and Appendices D and E. University of Illinois, *Chemistry: 1941–1951* (Urbana, Illinois, 1951).

6. D. Stanley Tarbell and Ann Tracy Tarbell, *Roger Adams: Scientist and Statesman* (Washington, D.C., 1981). Roger Adams, "The Relation of the University Scientists to the Chemical Industries," *Industrial and Engineering Chemistry, News Edition* 13 (1935), 365–367. Adams, "Universities and Industry in Science: Perkin Medal Address," *Industrial and Engineering Chemistry* 46 (1954), 506–510. D. S. Tarbell, Ann T. Tarbell, and R. M. Joyce, "The Students of Ira Remsen and Roger Adams," *Isis* 71 (1980), 620–626.

7. "Carl Shipp (Speed) Marvel," in Peter J. T. Morris, *Polymer Pioneers* (Philadelphia: CHOC, 1986), 61–63. Herman Mark, "The Contribution of Carl (Speed) Marvel to Polymer Science," *Journal of Macromolecular Science—Chemistry A21* (1984), 1567–1606. Burton C. Anderson, " 'Speed' Marvel at Du Pont," *Journal of Macromolecular Chemistry—Chemistry* A21 (1984), 1665–1687. Interview of Carl Marvel by Leon Gortler and Charles Price for the Beckman Center oral history program, 13 July 1983.

8. F. E. Matthews and H. M. Elder, British Patent 11,636, 1914.

9. D. S. Frederick, C. S. Marvel, and H. D. Cogan, *Journal of the American Chemical Society* 56 (1934), 1815–1819.

10. H. Staudinger, M. Brunner, and W. Feisst, *Helvetica Chimica Acta* 13 (1930), 805–832.

11. C. S. Marvel, J. H. Sample, and M. F. Roy, *Journal of the American Chemical Society* 61 (1939), 3241–3244.

12. Roger Adams, R. C. Fuson, and C. S. Marvel, "The Graduate Training of Chemists," in American Chemical Society, *Careers in Chemistry and Chemical Engineering* (Washington, D.C., 1951), 31–33. Interview of Raymond L. Myers by Peter Morris, 12 December 1985.

13. See monthly reports for August 1943 (CR 140), September 1943 (CR 175), and December 1943 (CR 240).

14. See monthly reports for December 1943 (CR 240), January 1944 (CR 252),

and March 1944 (CR 313). C. S. Marvel, G. E. Inskeep, P. V. Smith, and R. L. Frank, "Effects of Impurities in Butadiene and Styrene on GR-S Polymerizations," CR 265, 13/14 March 1944; R. L. Frank, J. R. Blegen, and P. V. Smith, "Butenes and Pentenes as Impurities in the Mutual Recipe," CR 398, 4 August 1944; R. L. Frank, Clark E. Adams, J. R. Blegen, Rudolph Deanin, and P. V. Smith, *Industrial and Engineering Chemistry* 39 (1947), 887–893; R. L. Frank, J. R. Blegen, G. E. Inskeep, and P. V. Smith, *Industrial and Engineering Chemistry* 39 (1947), 893–895.

15. C. S. Fuller and R. R. Williams, "Summary of Present Research on Synthetic Rubber," 15.

16. C. S. Marvel and Henry E. Baumgarten, "Chemical Study of the Structure of Diene Polymers and Copolymers," in Whitby, Davis, and Dunbrook (eds.), *Synthetic Rubber*, 294.

17. C. D. Harries, *Annalen der Chemie* 343 (1905), 311–375.

18. See monthly reports for September 1943 (CR 175) and January 1944 (CR 252). The September report indicates that these results were checked against those from infra-red spectrophotometry. J. K. Lawson, Jr., N. Rabjohn, and C. S. Marvel, "The Determination of Vinyl Side Chains in GR-S and Related Polymers by Ozonolysis," CR 234, 23 December 1943. N. Rabjohn, C. E. Bryan, G. E. Inskeep, H. W. Johnson, and J. K. Lawson, *Journal of the American Chemical Society* 69 (1947), 314–319.

19. R. Pummerer and P. A. Burkard, *Berichte* 55 (1922), 3458–3472.

20. I. M. Kolthoff and T. S. Lee, "Report LXXXIII: The Use of Perbenzoic Acid in the Analysis of Unsaturated Compounds. II. Determination of External Double Bonds in Synthetic Rubbers," CR 1075, 16 April 1946; Kolthoff, Lee, and M. A. Mairs, "Report XCVI: The Use of Perbenzoic Acid in the Analysis of Unsaturated Substances. III. Results of Determinations of External Double Bonds in Synthetic Rubbers," CR 1254, 11 October 1946. Kolthoff, Lee, and Mairs, *Journal of Polymer Science* 2 (1947), 199–228.

21. H. Rabjohn, C. S. Marvel, and C. E. Bryan, "The Investigation of the Structure of GR-S. Ozonolysis of Sample G–1," CR 146, 13/14 September 1943. Bryan, "The Structure of GR-S. Ozonolysis of Samples G–1 and G–23," CR 314, 29 April 1944. Rabjohn, Bryan, G. E. Inskeep, H. W. Johnson, and J. K. Lawson, *Journal of the American Chemical Society* 69 (1947), 314–319.

22. I. M. Kolthoff, T. S. Lee, and C. W. Carr, "Report LXXXV: Determination of Polystyrene in GR-S Rubber," CR 1093, 3 May 1946. Kolthoff, Lee, and Carr, *Journal of Polymer Science* 1 (1946), 429–433.

23. C. S. Marvel and R. E. Light, *Journal of the American Chemical Society* 72 (1950), 3887–3891.

24. *Ibid.*, 3888.

25. Fuller and Williams, "Summary of Present Research on Synthetic Rubber," 16.

26. D'Ianni, "Fun and Frustrations with Synthetic Rubber," G68.

27. The monthly report for February 1944 notes, "A program of making new monomers related to styrene . . . has been started" (CR 298, 3).

28. See monthly reports for February 1944 (CR 298) and March 1944 (CR 313).

29. J. D. D'Ianni, J. Marsh, and W. Findt, "Dimethyl Vinylethynyl Carbinol and 2-Vinylpyridine in Butadiene Copolymers and as Third Components in the GR-S Polymerization System," CR 93, 24/25 June 1943. D'Ianni, "Fun and Frustrations with Synthetic Rubber," G69.

30. "The University's Synthetic Rubber Program During World War II," November 1956, 8–10. The words outside the inner quotes are apparently a paraphrase of Marvel's remarks.

31. C. S. Marvel, et al., *Journal of Polymer Science* 8 (1952), 599–605. C. S.

Marvel, "University of Illinois Synthetic Rubber Research Related to Defense Requirements," *Proceedings, Joint Army-Navy-Air Force Conference on Elastomer Research and Development* (Washington, D.C., 1954), 54.

32. C. S. Marvel, et al., *Industrial and Engineering Chemistry* 45 (1953), 1532–1538.

33. C. S. Marvel, et al., *Industrial and Engineering Chemistry* 45 (1953), 2311–2317.

34. Marvel, "University of Illinois Synthetic Rubber Research Related to Defense Requirements," 55.

35. H. Mark and C. Wulff, German Patent 550,055, filed 9 August 1929.

36. Spitz, *Petrochemicals,* 184–187. Basil Achilladelis, "History of UOP: From Petroleum Refining to Petrochemicals," *Chemistry and Industry* (19 April 1975), 343.

37. Frederick M. Fowkes, "William Draper Harkins," in *Dictionary of Scientific Biography* (New York, 1972), volume 6, 117–119. George B. Kauffmann, "William Draper Harkins (1873–1951): A Controversial and Neglected American Physical Chemist," *Journal of Chemical Education* 62 (1985), 758–761. Warren C. Johnson, "William Draper Harkins," in Miles (ed.), *American Chemists and Chemical Engineers*. Additional information from C. S. Fuller.

38. Frank H. Westheimer, "Morris Selig Kharasch," *Biographical Memoirs of the National Academy of Sciences* 34 (1960), 123–134. Also see Frank R. Mayo, "The Evolution of Free Radical Chemistry at Chicago," *Journal of Chemical Education* 63 (1987), 97–99. The biography in W. A. Waters (ed.), *Vistas in Free Radical Chemistry,* page ix, is surprisingly brief, but see W. A. Waters and F. R. Mayo, "The Significance of the Work of M. S. Kharasch in the Development of Free-Radical Chemistry," in *ibid.,* 1–5; and (F. R. Mayo), "The Published Researches of M. S. Kharasch," in *ibid.,* 6–8.

39. Daniel P. Jones, "Julius Oscar Stieglitz," in Miles (ed.), *American Chemists and Chemical Engineers.*

40. M. S. Kharasch and F. W. Stavely, "The behavior of mercuric salts of organic acids towards heat," *Journal of the American Chemical Society* 45 (1923), 2961–2972.

41. M. S. Kharasch and R. Marker, "The decomposition of unsymmetrical mercuri-organic compounds: a method of establishing the relative degree of electronegativity of organic radicals," *Journal of the American Chemical Society* 48 (1926), 3130–3143.

42. M. S. Kharasch and F. R. Mayo, *Journal of the American Chemical Society* 55 (1933), 2468–2490. F. R. Mayo, "The Discovery of the Peroxide Effect," in W. A. Waters (ed.), *Vistas in Free Radical Chemistry,* 139–142; Mayo, "The Evolution of Free Radical Chemistry at Chicago," *Journal of Chemical Education* 63 (1987), 97–99.

43. M. S. Kharasch, H. Engelmann, and F. R. Mayo, *Journal of Organic Chemistry* 2 (1937), 288–302, 400, 577.

44. P. J. Flory, *Journal of the American Chemical Society* 59 (1937), 241–253. The *concept* of free radical polymerization is, of course, older. The first clear reference to free radical polymerization (of ethylene in the gas phase) appears in H. S. Taylor and W. H. Jones, *Journal of the American Chemical Society* 52 (1930), 1119–1120.

45. M. S. Kharasch, A. T. Read, and F. R. Mayo, *Chemistry and Industry* 57 (1938), 752. Kharasch's research for the rubber program is covered by Cheves Walling in two papers: C. Walling, "The Contributions of Morris S. Kharasch to Polymer Science," in W. A. Waters (ed.), *Vistas in Free Radical Chemistry* (London, 1959), 143–150; C. Walling, "The Development of Free Radical Chemistry," *Journal of Chemical Education* 63 (1986), 99–102.

46.   M. S. Kharasch and M. G. Berkman, *Journal of Organic Chemistry* 6 (1941), 810–817. Kharasch and F. H. Westheimer, "Oxygen Effect on Rate of Polymerization," CR 209, 6/7 December 1943. Kharasch, "Accelerators and Inhibitors of Copolymerization," CR 287, 13/14 March 1944.

47. M. S. Kharasch and F. H. Westheimer, "The Determination of the Role of Free Radicals in the Polymerization of Butadiene, in its Copolymerization with Styrene and in Similar Reactions," CR 77, 24/25 June 1943.

48. C. C. Price, R. W. Kell, and E. Krebs, *Journal of the American Chemical Society* 64 (1942), 1103–1106. Price and D. A. Durham, *Journal of the American Chemical Society* 65 (1943), 757–759.

49. R. B. Carlin, "A Mechanism for Butadiene-Styrene Copolymerization," CR 104, 16 July 1943.

50. For the research of Mayo and Walling at the University of Chicago and U.S. Rubber, see the interviews of Frank Mayo (21 January 1981) and Cheves Walling (12 September 1979) by Leon Gortler for the Beckman Center oral history program. Also see Cheves Walling, "Forty Years of Free Radicals," in W. A. Pryor (ed.), *Organic Free Radicals,* ACS Symposium Series 69 (Washington, D.C., 1978), 4–6.

51. F. R. Mayo, *Journal of the American Chemical Society* 65 (1943), 2324–2329.

52. A good summary of this work is provided by W. D. Harkins, *Journal of the American Chemical Society* 69 (1947), 1428–1444, and Fryling, "Emulsion Polymerization Systems," 233–238. Also see W. D. Harkins, "Report LVI: A General Theory of the Mechanism of Emulsion Polymerization II," CR 1824, 15 June 1948.

53. M. S. Kharasch, et al., "The Laboratory Control of Popcorn Polymer," CR 950, 14 January 1946. Kharasch, et al., *Industrial and Engineering Chemistry* 39 (1947), 830–837.

54. W. Nudenberg and M. S. Kharasch, "Emulsion Polymerization of Butadiene in the Light," CR 1654, 24 October 1947. Fryling, "Emulsion Polymerization Systems," 275–276.

55. M. L. Corrin, "A Kinetic Treatment of Emulsion Polymerization," CR 1326, 30 December 1946; Corrin, *Journal of Polymer Science* 2 (1947), 257–262.

56. W. V. Smith, *Journal of the American Chemical Society* 70 (1948), 3695–3702; W. V. Smith and R. H. Ewart, *Journal of Chemical Physics* 16 (1948), 592–599.

57. M. Morton, P. P. Salatiello, and H. Landfield, *Journal of Polymer Science* 8 (1952), 215–224.

58. M. S. Kharasch, A. Fono, and W. Nudenberg, *Journal of Organic Chemistry* 15 (1950), 753–762.

59. For two classic examples, see M. S. Kharasch, F. S. Arimoto, and W. Nudenberg, *Journal of Organic Chemistry* 16 (1951), 1556–1565; and Kharasch, F. Kawahara, and Nudenberg, *Journal of Organic Chemistry* 19 (1954), 1977–1990.

60. For general accounts of Kolthoff's career, see H. A. Laitinen and E. J. Meehan, "Happy Birthday I. M. Kolthoff," *Analytical Chemistry* 56 (1984), 248A–262A, and J. J. Lingane "Izaak Maurits Kolthoff," *Talanta* 11 (1964), 67–73. There are several interviews of Kolthoff: by George Tselos, for the Beckman Center oral history program, 15 March 1984; by Herbert Laitinen for the ACS Eminent Chemists Videotape Series; and by Robert Brasted for "Impact," *Journal of Chemical Education* 50 (1973), 663–666. There is also a set of fourteen tapes of untranscribed interviews of Kolthoff by Roger Steuer in 1980 at the University of Minnesota (copy stored at the Beckman Center for History of Chemistry).

61. Laitinen and Meehan, "Happy Birthday I. M. Kolthoff," 257A, states that Kruyt was the George Fisher Baker Lecturer in about 1926. Ironically, Ernst Cohen of the University of Utrecht—who was then on bad terms with Kolthoff—was the Baker Lecturer in 1926. The account given here is based on H. S. van Klooster, "Hugo Rudolph Kruyt," *Journal of Chemical Education* 19 (1942), 165; and the foreword to Harry Boyer Weiser (ed.), *Colloid Symposium Monograph,* volume five (New York: Chemical Catalog Co., 1928). This account agrees better with Kolthoff's comments in Brasted, "Impact," 665.

62. Brasted, "Impact," 665. Lingane, "Kolthoff," 69, says that Kolthoff's research on polarography began in the fall of 1934.

63. Brasted, "Impact," 665.

64. I. M. Kolthoff and W. E. Harris, "Preliminary Report III: The Determination of Mercaptan in Latex by Amperometric Titration with Silver Nitrate in Ammoniacal Alcoholic Medium," CR 62, 24/25 June 1943. Kolthoff and Harris, *Industrial and Engineering Chemistry, Analytical Edition* 18 (1946), 161–162.

65. I. M. Kolthoff and L. S. Guss, "The Determination of Persulfate in Latex," CR 249, 21 February 1944. Kolthoff, Guss, D. R. May, and A. I. Medalia, *Journal of Polymer Science* 1 (1946), 340–352.

66. I. M. Kolthoff and W. J. Dale, "Report XXXV: Contributions to the Mechanism of Emulsion Polymerizations. I. The Emulsion Polymerization of Styrene," CR 422, 7 September 1944. Kolthoff and Dale, *Journal of the American Chemical Society* 67 (1945), 1672–1674.

67. C. C. Price and C. E. Adams, "Kinetics of the Emulsion Polymerization of Styrene," CR 501, 14 December 1944. Price and Adams, *Journal of the American Chemical Society* 67 (1945), 1674–1680.

68. I. M. Kolthoff and W. E. Harris, "Report XXVII: A Comparision Between Primary and Tertiary Mercaptans as Modifiers and Activators in the Mutual Recipe," CR 355, 12/13 June 1944. Kolthoff and Harris, *Journal of Polymer Science* 2 (1947), 41–89.

69. I. M. Kolthoff, D. E. Williams, L. S. Guss, and C. W. Carr, "Report XXV: Studies of the Effect of Soap in the GR-S Polymerization. I. Effect of the Amount of Soap on the Rate of Conversion and the Properties of the Polymer," CR 337, 12 June 1944. Kolthoff, D. E. Williams, and C. W. Carr, "Studies on the Effect of Soap on the GR-S Polymerization. V. Polymerizations Started at 70 C with Varying Amounts of Soap and Different Mercaptans," CR 513, 5 December 1944. Kolthoff and E. J. Meehan, "Progress Report—January 1947," CR 1355, 63–71.

70. I. M. Kolthoff and L. S. Guss, "Report XIV: Effect of Soap Composition on GR-S Polymerization," CR 208, 6 December 1943. Kolthoff, D. E. Williams, and C. W. Carr, "Report XXXIX: Studies of the Effect of Soap on GR-S Polymerization. III. The Preparation of Well-Modified GR-S With Less Than the Mutual Recipe Amount of Soap," CR 459, 25 October 1944.

71. E. J. Meehan and I. M. Kolthoff, "Report V: Determination of Styrene by Ultraviolet Spectrophotometry," CR 117, 13 August 1943. Meehan, "Report XXVI: The Spectrophotometric Determination of Bound Styrene in GR-S Copolymers," CR 354, 12/13 June 1944. Meehan, "Report XXXVIII: The Effect of Polymerization Conditions and of Variations in the Recipe on the Styrene Content of Butadiene-Styrene Copolymers. The Styrene Content of Commercial Samples," CR 426, 14/15 September 1944. Meehan, "Report LIII: Relation Between the Styrene Content of GR-S and the Kinetics of the Polymerization Process. The Styrene Content of Polymer Increments. The Effect of Increment Monomer Addition Upon Polymer Homogenity," CR 617, 28 February 1945. Meehan, *Journal of Polymer Science* 1 (1946), 175–182 and 318–328.

72. W. B. Reynolds, "GR-S Polymerization: Diazo Thio-Ethers as Catalyst Modifiers," CR 408, 4 June 1944. Reynolds and E. W. Cotten, *Industrial and Engineering Chemistry* 42 (1950), 1905–1910. Reynolds and Cotten, U.S. Patent 2,501,692, filed 17 January 1946. I. M. Kolthoff and W. J. Dale, "Report XLIX: p-Methoxy Phenyl Diazo Thio-(2-Naphthyl)-Ether (Denoted as M.D.N.) as Activator-Modifier in Polymerizations by Various Recipes," CR 522, 27 December 1944. Kolthoff and Dale, *Journal of Polymer Science* 3 (1948), 400–409; Kolthoff and Dale, *Journal of Polymer Science* 5 (1950), 301–306.

73. C. F. Fryling, S. H. Landes, W. M. St. John, and C. A. Uraneck, *Industrial*

and Engineering Chemistry 41 (1949), 986–991. F. A. Bovey, I. M. Kolthoff, A. I. Medalia, and E. J. Meehan, *Emulsion Polymerization,* 360–369, 374–390. Frying, "Emulsion Polymerization Systems," 261–268.

74. I. M. Kolthoff and A. I. Medalia, *Journal of Polymer Science* 6 (1951), 209–223. Bovey, Kolthoff, Medalia, and Meehan, *Emulsion Polymerization,* 369–372.

75. G. S. Whitby, N. Wellman, V. M. Floutz, and H. L. Stephens, *Rubber Age* 65 (1949), 545; and *Industrial and Engineering Chemistry* 42 (1950), 445–456. Bovey, Kolthoff, Medalia, and Meehan, *Emulsion Polymerization,* 390–395. Fryling, "Emulsion Polymerization Systems," 272–274.

76. H. E. Albert, G. W. Gottschalk, and G. E. P. Smith, *Industrial and Engineering Chemistry* 40 (1948), 482–487. J. O. Cole, C. R. Parks, and J. D. D'Ianni, *Industrial and Engineering Chemistry* 42 (1950), 2553–2557.

77. J. M. Mitchell, R. Spolsky, and H. L. Williams, *Industrial and Engineering Chemistry* 41 (1949), 1592–1603. Rowzee and Buckler interview. Buckler, "Canadian Contribution to Synthetic Rubber Technology," 6–7.

78. I. M. Kolthoff and E. J. Meehan, *Journal of Polymer Science* 9 (1952), 433–452. Bovey, Kolthoff, Medalia, and Meehan, *Emulsion Polymerization,* 395–399.

79. Much of this research is covered by the monograph-sized (177 pp. typescript) report by A. I. Medalia and I. M. Kolthoff, "Free Radical Formation and Induced Reactions Accompanying the Reaction Between Ferrous Iron and Hydrogen Peroxide or Organic Hydroperoxides," CR 1858, 2 July 1948.

80. Henry John Horstman Fenton (1854–1929), while he was still an undergraduate at Christ's College, Cambridge, reported that a mixture of hydrogen peroxide and ferrous sulfate give an unusual violet coloration when added to a solution of tartaric acid. Five years later, he suggested that the color was a result of the oxidation of the tartaric acid rather than the iron salt. See H. J. H. Fenton, *Chemical News* (5 May 1876), 190; Fenton, *Chemical News* (11 March 1881), 110–111. Fenton was a University Demonstrator in Chemistry at Cambridge for many years and was elected a Fellow of the Royal Society in 1899. An obituary by W. H. Mills appeared in the *Journal of Chemical Society* 132 (1930), 889–894.

81. I. M. Kolthoff and A. I. Medalia, *Journal of the American Chemical Society* 71 (1949), 3789–3792.

82. I. M. Kolthoff and A. I. Medalia, *Analytical Chemistry* 23 (1951), 595–603.

83. Wyndham D. Miles, "Wilder Dwight Bancroft," in Miles (ed.), *American Chemists and Chemical Engineers.* John W. Servos, "A Disciplinary Program That Failed: Wilder D. Bancroft and the *Journal of Physical Chemistry* 1896–1933," *Isis* 73 (1982), 207–232. Also see Emile M. Chamot and Fred H. Rhodes, "The Development of the Department of Chemistry and of the School of Chemical Engineering at Cornell" (unpublished typescript, 1960), available at the Beckman Center for History of Chemistry.

84. Donald D. Fitts, "John Gamble Kirkwood," in W. D. Miles (ed.), *American Chemists and Chemical Engineers.*

85. For a good general account of Debye's work, see Charles P. Smyth, "Debye, Peter Joseph William," in C. C. Gillispie (ed.), *Dictionary of Scientific Biography,* volume 3 (New York, 1971), 617–621. Also see William O. Baker, "Peter Joseph William Debye," in W. O. Milligan (ed.), *Proceedings of the Robert A. Welch Conferences on Chemical Research,* volume 20, *American Chemistry—Bicentennial* (Houston, Texas, 1977), 154–199. H.-F. Eicke, "Peter J. W. Debye's Beiträge zur Makromolekularen Wissenschaft—ein Beispiel zukunftsweisender Forschung," *Chimia* 38 (1984), 347–353.

86. File on Peter Debye in Oliver Buckley's papers, Box 77, AT&T Archives, Warren, New Jersey. See especially the letter from James Lawrence Fly, Chairman of the Defense Communications Board, to O. E. Buckley, 30 January 1942, indicating that

Debye's previous clearance had been cancelled, partly on the grounds "that he is in some quarters regarded as an opportunist, . . . said to have 'conservative leanings' and . . . his scientific career was apparently wholly in Germany." Also see a letter to Buckley dated 16 March 1942 protesting the barring of Debye from Bell Laboratories, signed by fifteen members of the staff, including C. S. Fuller and W. O. Baker; and a letter from Paul A. Porter, Chairman of the Board of War Communications, to Buckley, 30 June 1945, restoring Debye's clearance following an appeal by Buckley, 18 May 1945.

87. Baker, "Debye," 189.

88. Hon. J. W. Strutt, *Philosophical Magazine* 4 (1871), 107–120. Milton Kerker, "Classics and Classicists of Colloid and Interface Science. 2. John William Strutt, Lord Rayleigh," *Journal of Colloid and Interface Science* 113 (1986), 589–593.

89. S. D. Gehman and J. E. Field, *Industrial and Engineering Chemistry* 29 (1937), 793–799. Gehman interview. Two chemists at the University of Louvain published similar research on proteins, see P. Putzeys and J. Brosteaux, *Transactions of the Faraday Society* 31 (1935), 1314–1325.

90. A. Einstein, *Annalen der Physik,* fourth series, 33 (1910), 1275–1298.

91. P. Debye, "Light Scattering in Solutions," CR 103, and "Scattering of Light in Solutions," CR 103A, 24/25 June 1943; P. Debye, *Journal of Applied Physics* 15 (1944), 338–342. Also see P. Debye, *Journal of Physical and Colloid Chemistry* 51 (1947), 18–32. These papers are reprinted in D. McIntyre and F. Gornick (eds.), *Light Scattering from Dilute Polymer Solutions* (New York, 1964). There is an excellent account of Debye's wartime activities in Baker, "Debye," 189–194.

92. P. Debye and E. S. Elyash, "Depolarization in Diluted Solutions," CR 110, 2 August 1943; recirculated as General Report 7 at the beginning of 1944. This report is reprinted in D. McIntyre and F. Gornick (eds.), *Light Scattering from Dilute Polymer Solutions.*

93. F. W. Billmeyer, Jr., "Particle Size of Polyvinyl-Chloride Molecular Weight 100,000," CR 497, 9 December 1944.

94. R. H. Ewart, C. P. Roe, P. Debye, and J. R. McCartney, "The Determination of Polymeric Molecular Weights By Light Scattering in Solvents-Precipitant Systems," CR 1139, 25 June 1946. Ewart, Roe, Debye, and McCartney, *Journal of Chemical Physics* 14 (1946), 687–695.

95. P. Debye and A. M. Bueche, *Journal of Applied Physics* 20 (1949), 518–525. P. Debye and A. M. Bueche, *Colloid Chemistry* 7 (1950), 33–46.

96. P. Debye and F. Bueche, *Journal of Chemical Physics* 19 (1951), 589–594.

97. W. H. Cashin, *Journal of Colloid Science* 6 (1951), 271–273.

98. P. Debye, *Journal of Physical and Colloid Chemistry* 53 (1949), 1–8. P. Debye and E. W. Anacker, *Journal of Physical and Colloid Chemistry* 55 (1951), 644–655.

99. F. Bueche, W. M. Cashin, and P. Debye, *Journal of Chemical Physics* 20 (1952), 1956–1958.

100. P. Debye and A. M. Bueche, "The Diffusion and Sedimentation Constants of Polymers," CR 1336, 17 January 1947. Debye and A. M. Bueche, *Science* 106 (1947), 507–508.

101. In polymer science, there are three kinds of molecular weights: (1) *number average molecular weight:* this is the simple average of the molecular weights of the individual molecular chains, and is obtained from osmotic pressure measurements; (2) *weight average molecular weight:* this gives greater value to the high molecular weight species and is calculated by dividing the sum of the square of the individual molecular weights by the sum of the molecular weights. This is obtained from light scattering measurements and is usually two or three times higher than the number average molecular weight; (3) *viscosity average molecular weight:* this is the molecular weight ob-

tained from viscosity measurements, and is usually less than weight average, depending on the value of a. I am grateful to James D'Ianni for his assistance with these definitions.

102. H. Staudinger and W. Heuer, *Berichte* 63 (1930), 222–234.

103. W. Kuhn, *Kolloid Zeitschrift* 68 (1934), 2–15.

104. P. Debye, "The Intrinsic Viscosity of Polymer Solutions," CR 1231, 14 October 1946: P. Debye and A. M. Bueche, *Journal of Chemical Physics* 16 (1948), 573–579.

105. P. J. Flory, *Journal of Chemical Physics* 17 (1949), 303–310. T. G. Fox, Jr., and Flory, *Journal of Physical and Colloid Chemistry* 53 (1949), 197–212. Flory and Fox, *Journal of the American Chemical Society* 73 (1951), 1904–1908. For experimental proof of ideal behavior in theta solvents, see W. R. Krigbaum, L. Mandelkern, and P. J. Flory, *Journal of Polymer Science* 9 (1952), 381–384.

106. T G. Fox, Jr., and P. J. Flory, *Journal of the American Chemical Society* 73 (1951), 1915–1920.

107. H. L. Wagner and P. J. Flory, *Journal of the American Chemical Society* 74 (1952), 195–200. P. J. Flory, L. Mandelkern, J. B. Kinsinger, and W. B. Shultz, *Journal of the American Chemical Society* 74 (1952), 3364–3367.

108. T. G. Fox, Jr., and P. J. Flory, *Journal of the American Chemical Society* 73 (1951), 1909–1915. W. R. Krigbaum and P. J. Flory, *Journal of Polymer Science* 11 (1953), 37–51.

109. L. Mandelkern, W. R. Krigbaum, H. A. Scheraga, and P. J. Flory, *Journal of Chemical Physics* 20 (1952), 1392–1397.

110. L. Mandelkern and P. J. Flory, *Journal of Chemical Physics* 10 (1951), 984–985.

111. Fuller and Williams, "Summary of Present Research on Synthetic Rubber," 17.

112. The minutes of this meeting were circulated as CR 2. The minutes of the next meeting of the PRDG (22/23 March 1943) were circulated as CR 46. Copies of the minutes of the ninth meeting (11–13 January 1945) and the tenth meeting (10–12 May 1945) were found in casefile 24060–1, volume C, AT&T Archives, Warren, New Jersey. No other minutes have been found.

113. C. S. Fuller, "Report of Polymer Research Branch to the Research Board [meeting] June 2, 1943," RG 234, RRC, Entry 231, PI–173, Minutes of the Rubber Research Board, 1.

114. "Rubber Division's Past Chairmen," in *Rubber Division 75th Anniversary* (Akron, 1984), 48.

115. Information from C. S. Fuller, 24 July 1987. It has also been suggested that the genteel members of Gibson Island Country Club took advantage of the club's improved financial position, after the war, to "blackball" the conferences because of the colorful behavior of certain polymer chemists.

116. Donald R. Hays, "Neil Elbridge Gordon," in W. D. Miles (ed.), *American Chemists and Chemical Engineers.*

117. Beatty interview. Also see McMillan, *The Chain Straighteners,* 19 and 23.

118. For the background to Ziegler and Natta's innovations, see Morris, *Polymer Pioneers;* and McMillan, *The Chain Straighteners,* 27–140. Strictly speaking, Ziegler worked at the Max Planck Institute for Coal Research, not a university, but the German Max Planck Institutes are closer in ethos to universities than most American research institutes. Charles Price's innovation was covered by U.S. Patent 2,866,744, issued 30 December 1958, but based on research carried out in 1949. C. C. Price, "How Chemists Create a New Product," *The Chemist* 38 (1961), 131–132.

119. Melvin Kranzberg, "Science-Technology and Warfare: Action, Reaction and

Interaction in the Post-World War II Era," in M. D. Wright and L. J. Paszek (eds.), *Science, Technology and Warfare* (Washington, D.C., 1971), 146–151.

120. *Ibid.,* 151.

121. For example, see Hugh Aitken, *Syntony and Spark: The Origins of Radio* (New York, 1976); Aitken, *The Continuous Wave* (Princeton, New Jersey, 1985). Clayton R. Koppes, *JPL and the American Space Program: A History of the Jet Propulsion Laboratory* (New Haven, 1982).

122. W. L. Semon, "A Suggested Program of Problems Suitable for Academic Research," typescript, 2 November 1942, RG 234, RRC, Entry 234, PI–173, Correspondence of R. R. Williams, 1 and 10.

123. This conclusion was independently reached by J. Langrish; Project Sappho at the University of Sussex; and R. F. Rothwell. For an overview of their conclusions and further details see Nuala Swords-Isherwood, *The Process of Innovation* (British-North American Committee, 1984), 12–13.

124. Taken from interview of I. M. Kolthoff by George Tselos. Maurice Visscher was head of the physiology department at the University of Minnesota from 1936 to 1960; he was not connected with the rubber program.

125. Harlan Trumbull, progress report of the synthetic rubber research section for August 1943, in "Progress Digests," 1943, B. F. Goodrich Collection, Series H, University of Akron archives.

126. J. D. D'Ianni, F. J. Naples, and J. E. Field, *Industrial and Engineering Chemistry* 42 (1950), 95–102. D'Ianni interview.

127. This conclusion is based on a survey of 227 Ph.D.'s who participated in the rubber research program between 1943 and 1946, and a more detailed study of the careers of 152 participants who obtained their Ph.D.'s from 1934 through 1948. The names were obtained from CR 1 through CR 990, and the necessary biographical information was taken from the 8th through the 14th editions of *American Men of Science* (Lancaster, Pennsylvania: Science Press, 1949/New York and London: Bowker, 1979) and *Who's Who in Engineering* (New York: Lewis Historical Publishing, 1954).

128. This point is discussed in Peter J. T. Morris, "Buna S Versus GR-S: A Comparative Study of Industrial Research in Germany and the United States," a paper presented at the HSS-BSHS joint meeting in Manchester, U.K., July 1988.

Chapter 4
Quality Control and
Polymer Evaluation

Introduction

Quality control, the standardization of GR-S, and the evaluation of new rubbers were key features of the rubber program. Bradley Dewey had stressed the overriding need for a uniform rubber to Ray Dinsmore in 1943.[1] The plant laboratories obviously did not exist until the plants themselves were constructed, and in any case they were not sufficiently well equipped or staffed to carry out all the necessary testing. Apart from the testing performed routinely by the rubber companies' central laboratories, these important tasks were undertaken by three institutions, which were neither universities nor part of the rubber industry.

The chemical section of Bell Telephone Laboratories (BTL) had been concerned with quality control since its foundation in 1895 as a two-man laboratory for the routine analysis of materials used by American Bell Telephone Company (the forerunner of AT&T).[2] The pressing need for materials that met the increasing stringent demands of modern communications technology brought BTL into polymer science in the 1930s. By 1941, BTL had assembled an experienced high-polymer research group as a source of basic information for the development of plastics and rubbers that could be used in telecommunications. A. R. Kemp and Frank S. Malm pioneered advances in the technology of wire insulation and deep-sea cables.

AT&T and the Federal Government were probably the two largest consumers of materials in the United States in the postwar period. The National Bureau of Standards (NBS) was founded on 1 July 1901 as an agency for the development of engineering standards (a successor to the Office of Weights and Measures), and soon thereafter it assumed the responsibility for the testing of materials purchased by the Federal Government.[3] It also undertook regular evaluations of commercial products. In the mid-1930s, the bureau became interested in the physical properties of isoprene and natural rubber, and then synthetic rubber as a result of an unplanned visit to the I.G. Farben laboratory at Leverkusen. The NBS came under the Department of Commerce, and Jesse Jones (who was then Secretary of Commerce) drew the bureau into the synthetic rubber program.

In contrast to the other two institutions, the Government Laboratories in Akron was not founded until 2 February 1944.[4] It was set up by the Rubber Reserve Company to referee the testing of GR-S by the plant laboratories and to evaluate new polymers. It was administered by the University of Akron but was completely separate from George Stafford Whitby's research group. Like Leverkusen, which may have inspired it, it consisted of a chemical laboratory, a pilot plant, and a rubber evaluation laboratory equipped for rubber testing.

In this chapter, we examine the research carried out by these three laboratories and assess their contributions to the rubber program. Particular attention is given to the influence of different environments and traditions. Were BTL and the NBS—world leaders in their own fields—able to apply this expertise to the problems presented by the rubber program? Was the attempt to set up an applied research center for the rubber industry successful?

Bell Telephone Laboratories

When R. R. Williams joined Western Electric (the manufacturing arm of AT&T) in 1919, he hired Archie R. Kemp, who had just graduated from the California Institute of Technology. During the 1920s, they investigated the physical properties of gutta-percha. Gutta-percha is a nonelastic isomer of rubber that had been used to insulate underwater telegraph cable since it was introduced by Werner Siemens in the 1860s. An attempt to use these undersea cables for telephone transmissions shortly after World War I had failed because of the higher demands imposed on the insulation by the more complex telephone signals. Kemp developed a suitable undersea insulation by mixing purified rubber and gutta-percha, which he called "paragutta." It was successfully used to link Catalina Island with the California mainland. Kemp also developed the "ultra-accelerator," a chemical that rapidly vulcanized raw rubber. This fast curing permitted the extrusion and vulcanization of rubber-coated wire in one swift continuous operation.[5]

BTL was also concerned with the changes produced in natural rubber during processing. Kemp's group investigated how the compounding and vulcanization of rubber could be improved. While studying the effect of milling on rubber, Kemp and Henry Peters became interested in the relationship between the petroleum ether soluble fraction of rubber (the sol) and the insoluble gelatinous fraction (the gel). In the late 1930s, most rubber chemists assumed that both types always existed in rubber, the gel forming either a skin around particles of sol rubber in the latex or a rigid sponge-like structure with the sol residing in the interstices.[6] Kemp and Peters published a new method of separating the two fractions in 1939.[7] In a subsequent paper, Kemp and Peters related the sol and gel with the molecular weights as measured by the viscosity of the solutions.[8] They showed that milling produced a narrower range of molecular weights.

BTL was also interested in finding alternatives to cotton and paper wire insulation. Calvin S. Fuller, a Ph.D. of the University of Chicago who had joined BTL in 1930, was sent to Cornell in 1934 to attend the Baker lectures given by J. R. Katz of Amsterdam University to find out more about X-ray diffraction of polymers.[9] Through the work of W. H. Carothers at Du Pont, BTL also became aware of the possible use of polyesters and polyamides. Fuller and Charles Erickson started a program to determine the structure of these polymers using X-ray diffraction and to relate the physical properties to the structure.[10]

In May 1939, the team was strengthened by the addition of William O. Baker, who had taken his doctorate at Princeton under Charles P. Smyth. Baker had studied the dielectrical properties of long-chain (but nonpolymeric) organic compounds at Princeton and quickly applied the results of this work to polyamides and polyesters.[11] The group was particularly interested in the effect of introducing disorder in the structure, which led to a decrease in polarity and crystallinity.[12] Much effort was devoted to find polymers superior in electrical properties to Du Pont's Nylon 66.

The synthesis of the polymers was mainly carried out by Burnhard S. Biggs, a Ph.D. from the University of Texas who had been a school teacher before joining BTL in 1936. In the course of this work, he developed Paracon, an unusual synthetic rubber that was a copolymer of ethylene sebacate and ethylene succinate, cross-linked by chemical oxidation.[13] It was self-sealing and was not attacked by oil, but it was also expensive and attacked by moisture, steam, acids, and alkalis. Its uses were therefore somewhat limited and, in any event, BTL was hardly in the business of producing polymers on a large scale. The production side was handed over to Resinous Products and Chemical Company, a Philadelphia-based subsidiary of Rohm and Haas. Although some consumers expressed interest, Paracon (or Paraplex, as Resinous Products called it) was never a success.

One of the objectives of the polymer work was to find a nonpolar material that would stand the high fields generated by the new coaxial cables. Kemp visited Imperial Chemical Industries, Ltd. (ICI) in England in 1939 and came back with a sample of ICI's polyethylene, which was a long-chain nonpolar hydrocarbon. The new polymer looked so promising that a larger desk-sized amount was obtained from ICI in 1940, made into strips, and incorporated into nine miles of the Washington-Baltimore telephone cable. It worked well, but many years of research was necessary before polyethylene insulation came into general use. By the mid-1950s, it had displaced lead as the preferred sheathing for electrical and communications cables, thereby completing that line of research on deep-sea cables started by Williams and Kemp thirty years earlier.[14]

When the synthetic rubber program began in the fall of 1942, Williams and Fuller went to work in the Office of the Rubber Director, as we noted in Chapter 1. It is not surprising that BTL was one of the first laboratories invited to join the program in December 1942.

Kemp, with the help of Malm and Peters, investigated the compounding of natural and synthetic rubber, studying the effect of inorganic and organic fillers and accelerators. He produced four of the six rubber compounding (RC) research reports published by the rubber program in 1943 and early 1944. Kemp was also in charge of the chemical analysis of rubbers (including samples of Russian and German rubber) until this task was taken over in late 1944 by the National Bureau of Standards. As part of his analytical duties, he estimated the styrene content of GR-S by the classical iodine monochloride method.[15] Initially, he also carried the intrinsic viscosity measurements, but this work passed over to Baker's group by 1944 for reasons that will become evident.

Biggs, working with Baker, studied the action of modifiers on the course of polymerization. He looked at the issue of overmodified and undermodified rubber. Like Robert Frank's group at Illinois, Biggs tried to synthesize a new compound that would slowly decompose under the conditions of the polymerization, continually releasing the desired modifier. Biggs looked at lauryl isothiourea (LITU) and other isothioureas, which are hydrolyzed to thiols.[16]

Biggs's work was highly regarded by Carl Marvel, but Williams decided, in October 1943, to abandon this line of research and to concentrate on the quality control work being developed by Baker. He wrote to Marvel:

> We are compelled to revise somewhat our plans for further work on synthetic rubber at the Bell Laboratories. Mr. Ingmanson who was in charge of our rubber covered wire development has left our employ and we have been compelled to enlist Biggs to almost the full extent of his time on rubber technology in order to repair this gap in our staff.

Largely for this reason, . . . we are inclined to confine our future efforts to betterment of quality control along the lines on which Baker has been working.[17]

For Baker, quality control was the watchword from the outset.[18] He was interested in the use of physical methods to study the structure of synthetic rubber, thereby gaining a greater understanding of its properties and how they could be improved.[19] From the prewar work on polyamides, he was aware of the importance of a regular chain structure in producing a satisfactory synthetic rubber.

Baker's research on the styrene content of GR-S also stemmed from his conviction that all the variables in GR-S had to be kept constant by careful and scientific quality control. The need to analyze for the styrene content arises from the fact that the styrene content of the final polymer is not the same as the proportion (charge ratio) of styrene monomer in the polymerization mixture. While Kemp was using the classical iodine monochloride method to measure styrene content, Baker was developing a method that measured the refractive index of a solution of rubber employing an interferometer.[20]

There was some debate in the summer and fall of 1943 about which was the better method.[21] The interferometer method required a carefully purified solution and there were unexplained discrepancies to be overcome, but the simplicity and reliability of this method—in expert hands—won the day. Samples of GR-S were sent to BTL for styrene analysis from July 1943 through September 1944, when the work was handed over to the National Bureau of Standards.

The BTL group noted that the styrene content fell during the fall and winter of 1943, with the result that the Rubber Director ordered the styrene charge to be increased in December 1943.[22] Baker later characterized this step as "probably the first direct application of research results from the government program to plant operation."[23] Baker's group then concentrated on improving the consistency of the styrene content from one batch to the next, and between factories. By measuring the styrene content of the samples sent in by the plants and sending the results back, they brought the variation down from 5% (usually too little styrene) to less than 0.5%, to about 24% ± 0.2% in September 1944.[24] This not only was critical for uniform processing but also increased the tensile strength and the low-temperature reliability, both key factors for GR-S.

Moreover, Baker had discovered a new type of gel in the early months of 1943.[25] If the polymerization of GR-S continued beyond a certain point, an open ("loose") three-dimensional network was formed throughout the latex particle. This molecular-scale microgel differed significantly in its properties from the conventional rubber gel (macrogel) studied earlier by Kemp and Peters. Baker characterized microgel using dilute solution viscosity and light scattering. He recently described the discovery of microgel networks as "much the most useful outcome" of the wartime research program.[26] Baker also noted that the drying of the raw rubber in the factories was producing an increased gel content in the final product. He realized that the absence of proper control of the manufacture of the rubber—during the polymerization and the drying stage, and of the composition of the GR-S—was producing a "tight" rubber with undesirable properties. Baker was convinced that a low gel content was crucial to the production of a good rubber, and thus a quick, reliable method for estimating gel content was sorely needed.

In June 1943, the Office of the Rubber Director circulated the method of determining the amount of gel in GR-S developed by Baker and James W. Mullen II.[27] With this "Vistex" method, a rubber producer could keep a check on his rubber by measuring the dilute solution viscosity (DSV) and the gel content of the latex. By the end of September 1943, Baker had constructed an easily manufactured apparatus. Paul Flory was given one of the first sets made.[28] This new apparatus enabled Mullen to make a complete study of gel formation on site at Goodyear's Torrance plant in California in November 1943, in conjunction with the California Development Program.[29] The work of relating gel content to the physical properties of the rubber and to ease of processing was done in collaboration with George Vila of U.S. Rubber in the fall and winter of 1943–1944.[30] Vila later became Chief Executive Officer of U.S. Rubber (Uniroyal) between 1961 and 1975.

While Baker was working on DSV, he became convinced that the widely used "Mooney viscosity," developed by Melvin Mooney of U.S. Rubber, was misleading.[31] He promoted his improved DSV method, which removed gel and chemical additives present in the rubber before measuring the viscosity, contrasting the scientific "Bell viscosity" with the ambiguous "Mooney." He also improved the design of the standard viscometer, in collaboration with Paul Flory, and claimed that previous viscosity measurements were inaccurate because of the failure to allow for all the possible variables:

I think the exchange of solutions [between Flory and Baker] was indeed profitable, since it definitely points towards new variables in accurate measurements of dilute solution viscosities. Shades of Staudinger; where is the "revised kiloHermann?["] Fortunately it is just a "leetle" error.[32]

During the first half of 1944, Baker and his group took steps to expedite the transfer of their methods to the plant laboratories. They took measures to improve the reproducibility of the results and standard samples were sent to the plants. A rapid method of testing latex was developed, so that the polymerization could be monitored continually and the final viscosity of polymer predicted.

However, the plants had difficulties with the new method and there was much confusion. Even Charles W. Perry of the ORD's Copolymer Equipment Development Branch felt obliged to raise various issues with Mullen:

> In connection with your recent work at the Goodyear, Los Angeles plant, I would appreciate it very much if you would straighten out one or two points. There seems to be a question in some rubber research chemists' mind that a little gel is perhaps a good thing for the quality of the rubber. . . . It is my impression that gel formed in either reaction or drying will be broken down by the physical action of milling. . . . Is this correct?[33]

In his reply, Mullen stated that gel was "definitely harmful to most of the measured physical properties" but conceded that it might be necessary to permit the formation of 10% gel in order to obtain 90% of the rubber "with a desirable distribution of molecular species." He went on to say that

> While *certain types* [stress in original] of milling do break down gel, it has never been shown that the resulting soluble, but seldom linear, fragments make any contribution to polymer quality. Hot milling only makes a bad situation worse, larger quantities of a stronger gel being produced.[34]

The issue of hot milling continued to be troublesome. Many technologists had accepted the idea that hot milling or thermal plasticization made the rubber more processable. The German industry actually operated on this principle. Baker realized that any excessive heating produced a tough and very undesirable gel. He argued that the apparent virtue of hot milling was counterproductive. The longer polymer chains were converted into gel. This made the rubber easy to extrude, since the more viscous fractions had been removed and the gel particles were suspended in the syrupy sol. The increased processibility was, however, far outweighed by the deterioration of the rubber's physical properties. Cold milling, by contrast, actually eliminated gel.[35]

Baker's views did receive some support from researchers in the rubber industry. In 1945, Leland White and his colleagues at U.S. Rubber published a paper on the effect of gel in GR-S on its physical and processing properties.[36] Like Baker, they distinguished between a loose dispersible gel and tight gel. They also found that tight gel affected both physical properties and ease of mixing. They concluded that the formation of tight gel should be avoided unless ease of processing was paramount.

Nonetheless, the plants were often happy enough with hot milling and felt that Baker was creating unnecessary difficulties. The gel was not completely worthless; in hot GR-S it provided most of the high molecular weight materials that gave the rubber its strength. Furthermore, hot milling was being replaced by mastication in a hot Banbury mixer, a machine not unlike a dough mixer.[37] The advantages of the mixer outweighed those of cold milling, especially as the partial exclusion of air in the mixer reduced gel formation.

Baker's quality control methods were regarded as very elegant, but impossible to use in the average plant laboratory. Less exacting requirements for quality control in the rubber industry compared with the Bell System and the semiempirical nature of contemporary rubber testing were part of the problem.

The inability of plant personnel to use the new methods in a consistent manner frustrated the BTL team. Baker complained to Les Friedman of the Polymer Research Branch of the Rubber Reserve Company in November 1944:

We are under constant pressure for work on many other war projects in addition to synthetic rubber, and since the technical status of the RRC and ORD [the Office of the Rubber Director, which actually no longer existed] is so well established, we are going to have largely to turn our meager efforts elsewhere. . . . Further, I cannot understand why Goodyear-Torrance needs further attention when just a year ago they were thoroughly versed in sol-gel technique. The answer, of course, probably lies in personnel turn-over, but I do not think the RRC is obligated to train each new batch of recruits. . . . Also, you probably recall that Mr. Maher, of Canadian Synthetic, Sarnia, recently visited here, and became thoroughly familiar with sol-gel determinations. Has this experience proved insufficient?[38]

A memorandum on possible postwar projects written at the end of August 1944 marked the end of Baker's single-minded concentration on synthetic rubber.[39] Thereafter, he spent less time on synthetic rubber research and returned to the study of other polymers of interest to BTL, especially the possible use of polymers in semi-conductors or dielectrics. For instance, Baker had a discussion with William Shockley on parallels between polymers and elemental selenium in October 1944.[40] The synthetic rubber research carried out at BTL became largely derivative, following up topics generated by other laboratories. Baker used the concentrated solution viscosity of rubber to determine the amount of branching present. He was also interested in the separation of synthetic rubber into different molecular weight fractions in order to study differences in properties.

BTL's direct involvement in the rubber research program was due to end on 1 September 1945, but given the then low level of expenditure it was possible to extend the contract to the beginning of 1946 without increasing the total grant.[41] However, Baker continued to play an important if low-key role as a critic and advisor from the sidelines. He still read the Copolymer Research (CR) reports and occasionally sent the authors (and the research section in Washington) his comments and suggestions.[42] He sought to tighten up loose arguments and point out the fallacies in weak ones.

It is possible to mention only a few examples of Baker's correspondence in this period. He wrote to Morris Kharasch in December 1945 about the classification of gels.[43] In July 1946, he wrote to Rudolph Deanin, at the University of Illinois, about the implications of his research on low-temperature polymerization for gel formation.[44] In the spring of 1947, he found the time to write a long letter to Kolthoff, in an attempt to bridge a divergence of opinion between the two laboratories about the properties of gel.[45] Finally, just before the casefile was closed in July 1947, he sent Samuel Gehman at Goodyear an implicit criticism of his work on the reaction between silver salts and synthetic rubbers.[46]

National Bureau of Standards

On the day Germany marched into Poland, 1 September 1939, the director of the NBS sent the Department of Commerce a memorandum about the services the bureau could provide if the United States was drawn into war.[47] Among the services listed was the testing of crucial materials including rubber. The NBS had studied natural rubber almost since its foundation, and isoprene was added in the early 1930s.[48] Two researchers at the bureau, Norman Bekkedahl, a chemical engineer, and Lawrence Wood, a physicist, met I.G. Farben's Erich Konrad at an international conference in 1938 and, as a result of this meeting, made their unexpected trip to Leverkusen.[49] To assess what he had learned at Leverkusen, Wood made a literature survey on synthetic rubber, which was published as a "best-selling" NBS circular in 1940.[50]

This publication brought Wood and Bekkedahl to the attention of Jesse Jones, who was then Secretary of Commerce, and he asked them in the fall of 1940 if it was really possible to make synthetic rubber, perhaps hoping their answer would be negative, so he could concentrate on the natural rubber stockpile. To cover himself, he asked the two NBS researchers to sign a minute recording their positive assessment.[51] When the Baruch Committee was convened in August 1942, Archibald McPherson, the head of the NBS rubber section, and Wood were among the technical advisors employed by the committee. Wood has recalled that they "were given a little piece out in left field to look into Thiokol as a possibility for retreads."[52]

During this period of uncertainty, the bureau was also asked to test processes for making synthetic rubber that were submitted by patriotic (or excessively optimistic) inventors. The chairman of the War Production Board, Donald M. Nelson, later recalled one such inventor:

Like all the other inventors who came to see us, he had his sample with him—a gummy, rubbery, substance, done up in paper—and, again like all the others, he had a marvelous story.

It went like this: he distilled oil from vegetable refuse, from practically any kind of household garbage, and found that this oil had marvelous properties—if a piece of crude rubber were immersed in it the rubber would actually grow in size. In two days a piece of crude rubber so immersed would double in size; if kept in the oil for a week it would grow to four times its original proportions. It was a piece of rubber treated in this way which he gave to Newhall. The solution to our problem was at hand: simply make huge vats of this oil, douse in it all the rubber we had on hand, and our precious stockpile would quadruple in a week.

Newhall sent him to the Bureau of Standards, on whose hard-working scientists we inflicted all these "inventors." The scientists listened to the man's story, analyzed his rubber, and presently reported to us that the man had stumbled on a well known phenomenon. It was possible to produce the kind of oil described, and the rubber that was soaked in it did absorb the oil and swell, but the catch was that as soon as the oil was put in a dehydrator the oil came out and the rubber returned to its normal size. Naturally, the moment you tried to process the rubber the dehydration process took place. Consequently, the whole business was of no use whatever.[53]

Tire tests had been a long-standing activity of the NBS, initially to check the tires purchased by Federal agencies such as the postal service. In the late 1920s, the American rubber industry greatly increased its use of reclaimed rubber. This was regarded with considerable skepticism by the bureau. McPherson showed in 1931 that reclaimed rubber tires wore so badly that it was worthwhile to pay five times as much for virgin rubber; a report that did not sit well with the rubber companies.[54] It was therefore natural for the Federal Government to ask the NBS to test GR-S tires until this task was largely taken over by the Government Tire Test Fleet in 1943. In the early 1950s, in conjunction with the development of the heavy-duty rubber tire, Robert Stiehler and John Mandel improved the standard tire wear tests used by the fleet.[55] This improvement was chiefly in the statistical treatment of the results rather than in the machinery used. At this time, the NBS had its own small fleet of seven trucks for road testing.

The NBS carried out five types of research for the rubber program:

studies of the monomers (especially ways to determine their purity), determination of thermodynamic data, improvement of rubber test methods, analysis of synthetic rubber, and studies of the thermal degradation of polymers. While the collection of a large body of thermodynamic data by the leading thermochemist Frederick Rossini and his collaborators was of industrial and scientific value, it was relatively straightforward, so we will restrict ourselves to the last three categories.

Whereas its predecessor BTL used solutions to obtain the styrene content of GR-S, the NBS measured the refractive index of the solid rubber, a technique developed by Wood and Irving Madorsky.[56] A purified sample of GR-S was pressed into a thin sheet and its refractive index measured with a conventional Abbe refractometer. The variation in the index is greater with the solid, which removed the need to use an expensive Zeiss interferometric refractometer. The NBS method was also easier to carry out—for example, no precise temperature control was required. This ease of operation was necessary as the BTL work had shown the value of carrying out daily checks of the styrene content, an impossible task for a central agency. The NBS technique was much more acceptable to the plant technicians than the tricky BTL method.

The NBS and BTL were also initially connected by the standard sample program, but BTL left this program when it ceased to carry out the styrene analyses. Samples were taken from a "standard bale" made under carefully controlled conditions by a rotating list of qualified plants.[57] After standardization at the NBS, they were shipped to the plant laboratories to be used as controls. Series of experimental and other special rubbers, for use in laboratory tests, were kept centrally in refrigerated storage. A team of experts from the bureau and Rubber Reserve regularly visited the rubber plants to instruct the laboratory technicians and answer queries. In conjunction with Rubber Reserve's Subcommittee on Test Methods and the American Society for Testing and Materials (ASTM), the NBS developed a collection of sixteen samples, covering natural and synthetic rubber, for standardization of rubber tests in 1948.[58]

The Mooney viscometer provides another contrast between BTL and the NBS. While Baker was keen to replace Mooney viscosity, Rolla Taylor of the NBS collaborated with the rubber industry (including Melvin Mooney himself) to improve it.[59] He showed that the viscometer had to be operated carefully to produce accurate results. This collaboration soon led to the introduction of an improved model.

The behavior of vulcanized rubber under stress, the stress-strain curve, is the most widely used test in the rubber industry. It has a variety of uses, including monitoring the curing of the rubber and for quality control, to ensure that the rubber was compounded properly and to the desired specifications.[60] In 1943, Wood and F. L. Roth measured the relationship between stress and temperature for GR-S.[61] They found that the then standard equipment introduced several errors into the measurement. Collaboration between the machine's manufacturers, Scott Testers, Inc., the NBS, and Rubber Reserve led to the introduction of an improved tester in 1948.[62] This permitted routine but accurate determination of points on the stress-strain curve with less effort.

The expertise of the NBS was also put to good use in the development of the chemical analysis of GR-S. This included improved methods of determining three important parameters: the water, oxygen, and ash content.[63] In 1952, Frederic Linnig and his colleagues developed a complete scheme of chemical analysis for the constituents of GR-S, using spectrophotometry, refractive index measurements, and classical titration.[64]

The NBS also made a careful study of the thermal decomposition of polymers, using the newly available routine mass spectrometer. Leo Wall found that polymers that form stable free radicals decompose, for the most part, to the original monomers, a process later called "unzippering."[65] With Robert Simha, the noted Austro-American polymer chemist at New York University, he devised a kinetic scheme for this type of depolymerization—the reverse of free radical polymerization—in 1952.[66] Simha had spent two years after the war reorganizing the bureau's polymer research.[67] Samuel Madorsky concurrently studied the pyrolysis of several polymers and classified their degradation in terms of two extremes: random decomposition at the weakest bonds and unzippering. Polymers that formed stable free radicals, like polymethyl methacrylate or polystyrene, largely unzipped to the monomer, but polyethylene and hydrogenated polystyrene broke down in a random fashion. Natural rubber and GR-S displayed an intermediate behavior: unzippering, but with a tendency to produce molecules larger than the monomers.[68]

Government Laboratories

The initial staff of the Government Laboratories were drawn from military units, such as the Chemical Warfare Service, and consisted of thirty-eight officers. By 1953, the total workforce had grown to over 180.[69] The work they carried out, between 1944 and the end of 1955, can be divided into three categories: testing and test methods (in cooperation with the NBS), polymer evaluation including pilot plant manufacture, and specific research projects commissioned by Rubber Reserve. This last category included the study of a continuous cold rubber process, oil-extended rubber, and alternative peroxides to cumene hydroperoxide.

The research on test methods, carried out under the general supervision of the Subcommittee on Test Methods, covered a variety of topics, including the Scott stress-strain tester,[70] a puncture tester,[71] the behavior of the Goodrich Flexometer (which measured the temperature rise on flexing) with synthetic rubbers,[72] and the measurement of tire temperatures by the Government Tire Test Fleet.[73] Perhaps the most significant project in this category was the investigation of low-temperature test methods in the early 1950s by Arthur Helin and B. Labbe in collaboration with three other laboratories.[74] Earlier in the program, Harold Leeper of the Government Laboratories had attempted to use the Shore A hardness test for this purpose, but he obtained erratic data.[75] Helin and Labbe concluded that the two most satisfactory methods were temperature retraction (because it was simple) and the Gehman torsion technique. The torsion method, developed by Samuel Gehman at Goodyear during the war, has remained a popular test, to the surprise of its modest inventor.[76]

To fulfill its mandate to evaluate existing and new polymers for use in the rubber program (a task taken over from the NBS), the Government Laboratories made and tested numerous rubbers, including most of the copolymers created by Marvel's group at Illinois. To give one example, Robert Laundrie and Milton Feldon studied ten copolymers and three terpolymers for their oil/solvent resistance and low-temperature flexibility.[77] Unfortunately, none of the combinations evaluated was superior in both respects to the standard butadiene-acrylonitrile copolymer.

In collaboration with the Eastern Regional Research Laboratory of the United States Department of Agriculture (USDA), the Government Laboratories evaluated Lactoprene EV.[78] This was a copolymer of ethyl acrylate (95%) and chloroethyl vinyl ether (5%). It was easy to process and possessed outstanding resistance to oil, light, and oxygen. Its heat resistance was greatly superior to butadiene-acrylonitrile copolymer, but less than the silicone rubbers. The USDA was interested in its development, because the ethyl acrylate could be made from lactic acid derived from milk or molasses. These routes were, however, undercut by synthetic acrylic acid produced cheaply in large quantities by Rohm and Haas. Goodrich commercialized Lactoprene EV (as Hycar PA 21) on a modest scale in 1948.

Encountering processing problems with the high molecular weight Alfin rubbers, William Taft and Henry Goldsmith injected air into the hot Banbury mixer to break it down. The Government Laboratories found virtues in the Alfin and sodium rubbers—the sodium rubbers were recommended for wire and cable coating—but these favorable assessments did not result in their commercial development.[79]

The technique of forced air injection into a Banbury mixer to plasticize cold rubber was studied by the Government Laboratories in the early 1950s.[80] Mastication without air had no effect, but the injection of air produced rapid plasticization. Laundrie, Taft, and M. Reich argued that the properties of the resulting material compared favorably with the untreated rubber. By careful selection of the parameters involved (especially time), it was possible to obtain a more easily processed rubber without an unacceptable loss of its physical properties.

Perhaps the most important task entrusted to the Government Laboratories by Rubber Reserve was the development in 1948 of a continuous polymerization process for cold rubber on the pilot plant scale.[81] The laboratories were asked to lay down standard operating conditions and to distribute the rubber produced to the rubber companies for evaluation and processing tests. The Government Laboratories took the batch process operated by Copolymer at its Baton Rouge plant (see Chapter 2) as its starting point. It used the system of batch reactors in series, with new material going into the first reactor and the latex pumped out the last one, used by the Germans during the war and developed by Goodyear for hot rubber. The results of this investigation were confirmed by the operation of the full-scale process by Goodrich a few months later.

The Government Laboratories then collaborated with the Goodrich plant at Institute, West Virginia, on the development of a sugar-hydroperoxide formula for "rapid" hot rubber instead of cold rubber.[82] This process did not require any change in the operating procedure and increased capacity by 30%, an important consideration in the middle of the Korean War. Shortly afterward, the Government Laboratories also produced a recipe that produced cold rubber at $-29°C$ using the "peroxamine" system invented by Whitby.[83]

At the beginning of the Korean War, Rubber Reserve also asked the Government Laboratories to investigate alternatives to the cumene hydroperoxide (CHP) used in the Custom recipe, to cover a possible shortage of cumene.[84] Turpentine was an obvious candidate because 70% of the world supply of turpentine was produced in the United States. The Government Laboratories cooperated with the Naval Stores Research Division, which was seeking new outlets for turpentine. Oxidized crude turpentine, without further refinement, was as effective as CHP. Furthermore, peroxides made from the pure hydrocarbons present in turpentine were superior to CHP. In collaboration with the regional laboratories of the USDA, the Government Laboratories also showed that methyl oleate peroxide (which can be produced from soybeans) was twice as active as CHP.[85]

Generic Research in the Rubber Program

Unencumbered by a dislike of government intervention or by an ignorance of rubber chemistry and technology, these institutions made a significant contribution to the rubber program. The strategy of improving GR-S through strict quality control and standardization was one of the rubber program's clear successes. It is not unimportant, perhaps, that the two founders of the research section, R. R. Williams and Calvin Fuller, were research leaders at BTL. The very high specifications necessary in the telephone industry had made quality control a crucial feature for BTL's research into new materials. The other activities covered in this chapter were also productive. This suggests that government-funded programs may be most effective when promoting "gray-area" basic research, which is neither pure academic research nor industrial research and development.

Nevertheless, the NBS was the best-placed of the three organizations to assist the rubber program. It had been associated with study and testing of natural rubber for over thirty years and had built up considerable expertise in this field. Its high standing and its position in the Federal administration ensured that it was taken seriously by the rubber industry. The NBS introduced an easily used but accurate method of determining the styrene content of GR-S. It helped to train the technicians in the plant laboratories and maintained a system of standard rubber samples. The NBS staff assisted with the development of improved testing equipment and produced a collection of thermodynamic and other physical data relating to GR-S and its monomers.

BTL was not so fortunate. Although its chemical section had studied rubber for almost as long as the NBS, its research in this field naturally excluded the important tire technology. While Kemp's group made contributions to the compounding of synthetic rubber, this research was overshadowed by the broader work in the rubber companies, which was more closely geared to the exigencies faced by the individual plants. Baker's interferometric method of styrene analysis materially assisted the program in the crucial period of 1943 and early 1944. Unsuited for use by the plant laboratories, it then faded away. The failure of the BTL group to fully understand the conditions under which the plants operated—different from those obtaining in the telecommunication industry—limited the contribution it made. This is particularly true of Baker's work on gel (and his rejection of Mooney viscosity). He failed to appreciate the importance rubber technicians placed on ease of processing, and they resented the strictures of someone outside the industry. It is perhaps not wholly coincidental that R. P. Dinsmore and Robert Juve cited White, not Baker, on the effect of tight gel in their essay on the processing of GR-S in *Synthetic Rubber*.

The Government Laboratories is best described as a qualified success. It appears to have been effective at the tasks assigned to it, including the evaluation of new rubbers, cold rubber, and oil-extended rubber. The laboratories also carried out an excellent cooperative survey of low-temperature testing. It was able to collaborate with the independently minded Goodrich on the development of a rapid hot rubber. Nonetheless, given the range of activities it encompassed and the extensive facilities at its disposal, the Government Laboratories had only a limited impact on the rubber program.

R. R. Nelson and R. N. Langlois have argued that the government can play a useful role in the funding of "generic" or "directed basic" research.[86] At first sight, a strong case can be made. It would not bring the government into competition with private firms. Indeed, the government would be funding research that might otherwise not receive sufficient funds, since it would be too general to be in the interest of any one company to fund it. However, the history of the Government Laboratories indicates a degree of caution. They were, in part, established as a generic research center to compensate the rubber companies for their loss of proprietary rights in the synthetic rubber field. Yet the Government Laboratories failed to command the support of the industry and were eventually sold to Firestone. It is not insignificant that the industry-dominated Polymer Research Policy Committee opposed their foundation.[87] In contrast to Britain, industry-wide trade association research laboratories were generally unpopular in the United States.[88] Most of the generic research in the American rubber industry is now carried out by the large rubber companies, and to a lesser extent by the University of Akron. This indicates that the success of government sponsorship of generic research will depend on the corporate and technical features of the industry involved. Government-sponsored generic research is likely to be more successful in technologically less advanced industries or in economic sectors with many small companies. It is not surprising that agriculture, generally regarded as one of the successes of government-sponsored generic research, originally belonged to both categories.

There is a remarkable parallel between the Government Laboratories and the failure of the pre–World War I campaign to establish a Chemische Reichsanstalt (Imperial Chemical Research Institute) in Germany. The American rubber industry refused to support the Government Laboratories after it was established. By contrast, the German government would not pay the running costs of the Reichsanstalt, although the chemical industry was eventually willing to donate most of the construction funds.[89]

It is interesting that the Government Laboratories did not develop a strategy to ensure its long-term survival, given the lack of any center of generic rubber research outside the rubber companies except the National Bureau of Standards. When NASA's Ames Research Center in California was threatened by government cutbacks on space research in the 1970s, the director of Ames, Hans Mark (the son of Herman Mark), persuaded NASA to adopt an "area of emphasis" strategy. By creating a particular research area that was unique to Ames, its "area of emphasis," Mark ensured that it survived the economy cuts.[90] It may simply be that there was no one of the necessary caliber at the helm of the Government Laboratories. Nonetheless, it is still remarkable that no sustained effort was made to preserve this institution and thereby challenge the near-monopoly of the British Rubber Producers' Research Association (now the Malaysian Rubber Producers' Research Association) in this area.

The administration of the laboratories by the University of Akron may have been an additional drawback. It is surprising that the Government Laboratories were not placed under the NBS, which was experienced in the fields covered by them, or directly administered by Rubber Reserve. Indeed, in 1947, the bureau's new director, Edward U. Condon, suggested that all government-sponsored rubber research be coordinated by the NBS.[91] These more effective modes of administration may have been prevented by pressure from the rubber companies. The employment of experienced polymer chemists and rubber technologists may have been prevented by the shortage of trained personnel, and the reluctance of chemists in the industry to abandon established or promising careers.

We have hitherto ignored the considerable efforts made in rubber research by the central laboratories of the rubber companies. Gehman's section in Goodyear was responsible for the development of a low-temperature tester, a semi-automatic dynamic tester, and the Instron tensile tester, which are still in use. The Goodrich Flexometer and the Firestone resonance apparatus have displayed similar stamina. The Mooney viscosimeter was introduced by U.S. Rubber, who (as we have seen) also carried out extensive research on the negative effects of gel.

As private companies, with reputations to maintain and customers to keep happy, the rubber companies had a vested interest in quality control. While the work of BTL, the NBS, and the Government Laboratories aided them in making the switch from natural rubber to synthetic rubber, they were well equipped to take charge of their own quality control once the plants were operating smoothly. Indeed, John Livingston and John Cox, two senior administrators in the rubber program, attributed the rapid increase in uniformity of the rubber from different plants to a mixture of wartime patriotism and competition between the plants (and companies) to produce the best grade.[92]

Notes

1. Memorandum from Bradley Dewey to R. P. Dinsmore, 5 April 1943. Appended to the minutes of the fourth meeting of the Rubber Research Board, RG 234, RRC, Entry 231, PI–173.

2. S. Millman (ed.), *A History of Engineering and Science in the Bell System. Physical Sciences (1925–1980)* (AT&T Bell Telephone Laboratories, 1983), 404.

3. Rexmond C. Cochrane, *Measures for Progress: A History of the National Bureau of Standards,* NBS Miscellaneous Publication 275 (Washington, D.C., 1966), reprinted by Arno Press (New York, 1976).

4. "Government Laboratories of the Office of Rubber Reserve," *Chemical and Engineering News* 24 (1946), 3024–3026. The Government Laboratories were originally envisaged by R. R. Williams as a means of compensating the rubber industry for "the extensive expropriation of private rights by government." Williams assumed that most of the long-term research would be concentrated at this new institution. He also saw the laboratories as a way of transferring the research program into the private sector, as "its control and the responsibility for its support should progressively pass into the hands of industry in proportion as government withdraws from rubber production." R. R. Williams, "Future Research on Synthetic Rubber," 30 June 1943, in casefile 24060–1, volume A, AT&T Archives, Warren, New Jersey (hereafter AT&T).

5. M. Fagen (ed.), *A History of Engineering and Science in the Bell System. The Early Years, 1875–1925* (Bell Telephone Laboratories, 1975), 986–987. Interview of A. R. Kemp for the oral history program of ACS Rubber Division, tape no. 15, 2 April 1965, no transcript, University of Akron archives. Millman, *Physical Sciences,* 518–519.

6. Harry Barron, *Modern Rubber Chemistry,* second edition (London, 1947), 91 and 105.

7. A. R. Kemp and H. Peters, *Journal of Physical Chemistry* 43 (1939), 923–939.

8. *Ibid.,* 1053–1082.

9. Private communication from C. S. Fuller to James Bohning, 30 June 1987.

10. Millman, *Physical Sciences,* 482–483. C. S. Fuller and C. L. Erickson, *Journal of the American Chemical Society* 59 (1937), 344–351. W. O. Baker, Fuller, and J. H. Heiss, Jr., *Journal of the American Chemical Society* 63 (1941), 2142–2148.

11. See interviews of William O. Baker by Jeffrey L. Sturchio and Marcy Goldstein for Beckman Center for History of Chemistry and AT&T oral history programs, 23 May 1985 and 18 June 1985.

12. W. O. Baker and C. S. Fuller, *Journal of the American Chemical Society* 64 (1942), 2399–2407. Baker and Fuller, *Journal of the American Chemical Society* 65 (1943), 1120–1130.

13. B. S. Biggs and C. S. Fuller, *Chemical and Engineering News* 21 (1943), 962–963. Biggs, *Bell Laboratories Record* 22 (1944), 317–319. Biggs, R. H. Erickson, and Fuller, *Industrial and Engineering Chemistry* 39 (1947), 1090–1097. The development of Paracon, such as it was, can be traced in casefile 37645-2, AT&T.

14. Millman, *Physical Sciences,* 483–484. Baker interview, 23 May 1985.

15. A. R. Kemp and H. Peters, "Determination of Styrene Content in GR-S Latex and Crude Polymer by the Iodine Chloride Method," CR 56, 15 September 1943. Unless otherwise noted, the details of the rubber research at BTL come from a careful reading

of the material in casefile 24060–1, volume A (December 1942 to December 1943), volume B (January to August 1944), and volume C (August 1944 to June 1945), AT&T. For a general survey, see C. S. Fuller, "Some Recent Contributions to Synthetic Rubber Research," *Bell System Technical Journal* 25 (1946), 351–384.

16. B. S. Biggs, W. S. Bishop, and W. J. Myles, "Alkylthioureas as Modifiers in the GR-S Reaction," CR 71, 24/25 June 1943. Biggs, Bishop, and R. H. Erickson, "Modifier Action and Locus of Reaction," CR 88, 24/25 June 1943. Biggs, Bishop, Erickson, and I. J. Gruntfest, "Further Experiments with Lauryl-isothiourea as a Modifier," CR 170, 13/14 September 1943.

17. R. R. Williams to C. S. Marvel, 20 October 1943, casefile 24060–1, volume A, AT&T.

18. See J. W. Mullen to C. W. Perry, Office of the Rubber Director, 26 November 1943, casefile 24060–1, volume A, AT&T: "The research program at the Laboratories under Dr. Baker has from the outset had the quality control of GR-S as its ultimate aim."

19. W. O. Baker, "Problems for a Physical-Organic Investigation of Synthetic Rubber—Case 24060," 7 December 1942, casefile 24060–1, volume A, AT&T. Also see C. S. Fuller and B. S. Biggs, "Linear Polyesters as Model Molecules for Rubber Research," CR 8, 19 December 1942.

20. W. O. Baker and J. H. Heiss, Jr., "Styrene Content and Optical Refraction of Butadiene Copolymers," CR 34, 23 March 1943. Baker and Heiss, "Styrene Content Variations in Butadiene Copolymers," CR 69, 24 June 1943.

21. See A. R. Kemp to J. D. Fennebresque, Copolymer Equipment Development Branch, 16 June 1943, and C. S. Fuller to C. W. Walton, Goodyear, 15 September 1943, casefile 24060–1, volume A, AT&T.

22. W. O. Baker, "Report of Work by Bell Telephone Laboratories, Inc., for Rubber Reserve Co., For Period Sept. 12, 1943 to Jan. 1, 1944," 21 March 1944, casefile 24060–1, volume B, AT&T, 5.

23. Baker interview, 18 June 1985.

24. W. O. Baker, "Report of Work by Bell Telephone Laboratories, Inc., for Rubber Reserve Co., For Period September 1, 1944, to October 1, 1944," [c. 29 November 1944], casefile 24060–1, volume C, AT&T. C. S. Fuller, "Some Recent Contributions to Synthetic Rubber Research," 366.

25. W. O. Baker, "Discovery and Application of Microgel Molecules in the National Synthetic Rubber Program," 13 October 1988. Copy in W. O. Baker Papers, AT&T. Baker and J. W. Mullen II, "Solubility Relationships in GR-S Polymer," CR 33, 23 March 1943.

26. Baker, "Microgel Molecules," 2–3.

27. W. O. Baker and J. W. Mullen II, "Sol-Gel Separation in GR-S Polymers," CR 70, 8 June 1943.

28. J. W. Mullen II and W. O. Baker, "Gel Content and Quality Control of GR-S," CR 148, and "Procedures for the Determination of Gel Content, Swelling Index, and Dilute Solution Viscosity of the Sol of GR-S Polymer," CR 148A, 13/14 September 1943. The large-scale production of the apparatus was undertaken by Otto Grainer Co., Newark, New Jersey, see Baker, "Report . . . Sept. 12, 1943 to Jan. 1, 1944," 1. W. O. Baker to P. J. Flory, Esso Laboratories, 23 September 1943, casefile 24060–1, volume A, AT&T.

29. J. W. Mullen II, "Research Aspects of Quality Control Measurements as Carried Out in a Standard Plant," CR 260, 13 March 1944 (this is the revised version of a report originally submitted on 30 November 1943).

30. W. O. Baker, "Report of Work by Bell Telephone Laboratories, Inc., for

Rubber Reserve Co., For Period Feb. 1, 1944 to Mar. 1, 1944," [March 1944], casefile 24060–1, volume B, AT&T, 1.

31. W. O. Baker to R. W. Kixmiller, 17 December 1943, casefile 24060–1, volume A, AT&T. W. O. Baker and J. W. Mullen II, "Examples of Control Measurements on GR-S Polymers: Dilute Solution Viscosities," CR 200, 6/7 November 1943.

32. Baker to Flory, Esso Laboratories, 24 August 1943, casefile 24060–1, volume A, AT&T.

33. Perry to Mullen, 8 November 1943, casefile 24060–1, volume A, AT&T.

34. Mullen to Perry, 12 November 1943, casefile 24060–1, volume A, AT&T.

35. W. O. Baker, "Report of Work by Bell Telephone Laboratories, Inc., for Rubber Reserve Co., For Period Feb. 1, 1944 to Mar. 1, 1944," [March 1944]; Baker, "Report of Work by Bell Telephone Laboratories, Inc., for Rubber Reserve Co., For Period March 1, 1944 to April 1, 1944," [April 1944], casefile 24060–1, volume B, AT&T. Baker, R. W. Walker, and N. R. Pape, "Interrelations of Sol and Gel in GR-S Polymers," CR 352, 16 June 1944.

36. L. M. White, et al., *Industrial and Engineering Chemistry* 37 (1945), 770–775.

37. Pierson interview.

38. W. O. Baker to L. A. Friedman, Jr., 29 November 1944, casefile 24060–1, volume C, AT&T.

39. W. O. Baker, "Post-War Projects," 25 August 1944, laboratory notebook 18940, February 1943 to April 1946, AT&T, 80. I am grateful for Jeffrey Sturchio's assistance with this matter.

40. "Discussion in morning (luncheon afterwards) with Messrs. W. Shockley, J. Shire, R. O. Grisdale on Se semi-conductors," 24 October 1944, *ibid.*, 177.

41. W. O. Baker in a memorandum to C. S. Fuller, (27 September 1945, casefile 24060–1, volume D, AT&T) suggested that the remaining $2,500 of RRC funds be used to maintain the BTL's relationship with the rubber research program, but without doing any active research.

42. W. O. Baker, "Comments on Report CR 865—'Cross-Linked GR-S for an Improved Processing Polymer,' " CR 865A, 25 October 1945, and Baker, "Comments on CR 873— 'Determination of Gel in GR-S,' " CR 873A, 25 October 1945.

43. W. O. Baker to M. S. Kharasch, 6 December 1945, casefile 24060–1, volume D, AT&T.

44. W. O. Baker to Rudolph Deanin, 15 July 1946, casefile 24060–1, volume D, AT&T.

45. W. O. Baker to I. M. Kolthoff, 14 March 1947, casefile 24060–1, volume D, AT&T.

46. W. O. Baker to S. D. Gehman, 22 July 1947, casefile 24060–1, volume D, AT&T.

47. Cochrane, *Measures for Progress*, 365–366, citing NBS Box 429, AG, Memo. LJB [riggs] to John M. Johnson, assistant to Secretary of Commerce, 1 September 1939.

48. Cochrane, *Measures for Progress*, 279–280. Lawrence A. Wood, "Memories of Synthetic Rubber at the National Bureau of Standards: 1936–1956," in *World War II Synthetic Rubber Program: Mission, Record, Mechanisms, Significance and Its Messages for Today* (typescript published by the Washington Rubber Group and Rubber Division, American Chemical Society, 1979), 23.

49. Wood, "Memories of Synthetic Rubber," 24–26. Cochrane, *Measures for Progress,* 411.

50. L. A. Wood, *Synthetic Rubbers: A Review of Their Composition, Properties and Uses,* NBS Circular 427 (Washington, D.C., 1940); reprinted in *Rubber Chemistry and Technology* 13 (1940), 861–885.

51. Wood, "Memories of Synthetic Rubber," 26–27.

52. *Ibid.*, 29.

53. Donald M. Nelson, *Arsenal of Democracy: The Story of American War Production* (New York, 1946), 300.

54. Cochrane, *Measures for Progress,* 280, citing A. T. McPherson, NBS Circular 393 (Washington, D.C., 1931), 17.

55. Review in *Rubber Age* 71 (1952), 361–364; also see R. Stiehler, "Development of a Tire Tester," in *Proceedings, Joint Army-Navy-Air Force Conference on Elastomer Research and Development* (Washington, D.C., 1954), 111–113. J. Mandel, M. N. Steel, and R. D. Stiehler, *Industrial and Engineering Chemistry* 43 (1951), 2901–2908. Mandel, Steel, and Stiehler, *Rubber Chemistry and Technology* 25 (1952), 656–686. R. D. Stiehler, G. G. Richey, and J. Mandel, *Rubber Age* 73 (1953), 201–208.

56. L. A. Friedman, Jr., "Comparison of Styrene Content of GR-S by Abbe and Interferometer Methods," CR 423, 14/15 September 1944. Irving Madorsky and L. A. Wood, "Measurement of Refractive Index and Determination of the Styrene Content of GR-S Copolymers," CR 442, 14/15 September 1944. Madorsky and Wood, "Procedure for the Measurement of the Refractive Index of Specification GR-S," CR 495, 30 November 1944. A. Arnold, Madorsky, and Wood, *Analytical Chemistry* 23 (1951), 1656–1659. L. A. Wood, "Physical Chemistry of Synthetic Rubbers," in Whitby, Davis, and Dunbrook (eds.), *Synthetic Rubber,* 323–324.

57. L. Meuser, R. D. Stiehler, and R. W. Hackett, *India Rubber World* 117 (1947), 57–61. Stiehler and Hackett, *Analytical Chemistry* 20 (1948), 292–296. Livingston and Cox, "Manufacture of GR-S," 206–207.

58. A. E. Juve, "Physical Test Methods and Polymer Evaluation," in Whitby, Davis, and Dunbrook (eds.), *Synthetic Rubber,* 507.

59. Rolla H. Taylor, *Factors Affecting Results Obtained With the Mooney Viscometer* NBS Circular 451 (Washington, D.C., 1945). Taylor, *Rubber Chemistry and Technology* 19 (1946), 808–820. Taylor, J. H. Fielding, and M. Mooney, *Rubber Age* 61 (1947), 567–573, 705–710, 738.

60. A. E. Juve, "Physical Testing," in Morton (ed.), *Introduction to Rubber Technology,* 469–475.

61. F. L. Roth and L. A. Wood, "Relations Between Stress, Strain and Temperature in a Pure-Gum Vulcanizate of GR-S," CR 61, 24/25 June 1943. Roth and Wood, *Journal of Applied Physics* 15 (1944), 749–757. Roth and Wood, *Rubber Chemistry and Technology* 18 (1945), 353–366.

62. W. L. Holt, E. C. Knox, and F. L. Roth, *Journal of Research of the National Bureau of Standards* 41 (1948), 95–102. Holt, Knox, and Roth, *Rubber World* 118 (1948), 513–517, 578.

63. Moisture: M. Tryon, *Journal of Research of the National Bureau of Standards* 45 (1950), 362–366. Oxygen: W. W. Walton, F. W. McCulloch, and W. H. Smith, *Journal of Research of the National Bureau of Standards* 40 (1948), 443–447. Ash: F. J. Linnig, L. T. Milliken, and R. I. Cohen, *Journal of Research of the National Bureau of Standards* 47 (1951), 135–138.

64. F. J. Linnig, J. M. Peterson, D. M. Edwards, and W. L. Acherman, *Analytical Chemistry* 25 (1953), 1511–1515; Linnig and A. Schneider, *Analytical Chemistry* 25 (1953), 1515–1517.

65. L. A. Wall, *Journal of Research of the National Bureau of Standards* 41 (1948), 315–322.

66. R. Simha, L. A. Wall, and P. J. Blatz, *Journal of Polymer Science* 5 (1950), 615–632.

67. Cochrane, *Measures for Progress,* 477–478.

68. S. L. Madorsky and Sidney Straus, *Journal of Research of the National Bureau of Standards* 40 (1948), 417–425. Madorsky, Straus, D. Thompson, and L. Williamson, *Journal of Research of the National Bureau of Standards* 42 (1949), 499–514. Madorsky, *Science* 111 (1950), 360–361. Madorsky, *Journal of Polymer Science* 9 (1952), 133–156. Madorsky, *Journal of Polymer Science* 11 (1953), 491–506. Straus and Madorsky, *Journal of Research of the National Bureau of Standards* 50 (1953), 165–176.

69. O'Callaghan, *The Government's Rubber Projects,* volume 2, 571.

70. D. R. Scheu and J. W. Schade, *India Rubber World* 112 (1945), 65–66.

71. H. M. Leeper, *Rubber Age* 59 (1946), 73–74.

72. B. G. Labbe, *India Rubber World* 128 (1953), 193–198.

73. V. J. Horning, *Rubber Age* 74 (1953), 395–396.

74. A. F. Helin and B. G. Labbe, *India Rubber World* 126 (1952), 227–231, 365–368.

75. H. M. Leeper, *India Rubber World* 115 (1946), 215–222. Leeper, *Rubber Chemistry and Technology* 20 (1947), 959–862.

76. Gehman interview.

77. R. W. Laundrie, M. Feldon, and A. L. Rodde, *India Rubber World* 122 (1950), 683–684.

78. W. M. Howerton, T. J. Dietz, A. D. Snyder, and G. E. Alden, *Rubber Age* 72 (1952), 353–362. C. H. Fisher, G. S. Whitby, and E. M. Beavers, "Miscellaneous Synthetic Elastomers," in Whitby, Davis, and Dunbrook (eds.), *Synthetic Rubber,* 900–910.

79. Alfin: W. K. Taft and H. Goldsmith, *Industrial and Engineering Chemistry* 42 (1950), 2542–2546. Sodium: M. H. Reich, R. E. Schneider, and W. K. Taft, *Industrial and Engineering Chemistry* 44 (1952), 2914–2922.

80. G. J. Tiger, M. H. Reich, and W. K. Taft, *Industrial and Engineering Chemistry* 42 (1950), 2562–2569. Reich, Taft, and R. W. Laundrie, *Rubber Age* 70 (1951), 55–62. Reich and Taft, *Rubber Age* 70 (1951), 55–62. Reich and Taft, *Rubber Age* 72 (1953), 619–624.

81. R. W. Laundrie and R. F. McCann, *Industrial and Engineering Chemistry* 41 (1949), 1568–1570. Laundrie, et al., *Industrial and Engineering Chemistry* 42 (1950), 1439–1442, 2358. M. Feldon, McCann, and Laundrie, *India Rubber World* 128 (1953), 51–53, 63.

82. Government Laboratories, *Chemical and Engineering News* 30 (1952), 136. R. F. McCann, R. W. Laundrie, D. Druesdow, and R. S. Reynolds, *India Rubber World* 129 (1953), 209–212, 214.

83. Fryling, "Emulsion Polymerization Systems," 274.

84. G. S. Fisher, L. A. Goldblatt, I. Kniel, and A. D. Snyder, *Industrial and Engineering Chemistry* 43 (1951), 671–674.

85. D. Swern, et al., *Journal of Polymer Science* 11 (1953), 487–490.

86. Richard R. Nelson and Richard N. Langlois, "Industrial Innovation Policy: Lessons from American History," *Science* 219 (1983), 814–818.

87. Private information from C. S. Fuller, citing a letter from R. R. Williams to Bradley Dewey, 1943.

88. Hounshell and Smith, *Science and Corporate Strategy,* 6.

89. Jeffrey A. Johnson, "The Chemical Reichsanstalt Association: Big Science in Imperial Germany" (Princeton University Ph.D. thesis, 1980); in revision as *The Kaiser's Chemists: The Modern Scientific Research Institutions in Imperial Germany* (Chapel Hill,

North Carolina, forthcoming). See pages 338–372 of the thesis, or Chapters Four and Five of the book.

90. Hans Mark and Arnold Levine, *The Management of Research Institutions: A Look at Government Laboratories* (Washington, D.C., 1984).

91. Cochrane, *Measures for Progress*, 442.

92: Livingston and Cox, "Manufacture of GR-S," 176.

Chapter 5
The Rubber Research Program and Polymer Science

Introduction

Polymer science expanded enormously in the United States between 1940 and 1965. New theories and techniques were developed. Academic courses and postgraduate degrees were established, initially at Brooklyn Polytechnic and the University of Chattanooga, and at several other universities by 1955. The *Journal of Polymer Science* was founded in 1946. Paul Flory published his classic monograph, *Principles of Polymer Chemistry* in 1953, which was followed by Fred Billmeyer's more elementary but enduring *Textbook of Polymer Chemistry* in 1957. The American Chemical Society's Division of Polymer Chemistry (originally the Division of High Polymer Chemistry) was established in 1950. The number of polymer chemists—as measured by the membership of this division—increased by leaps and bounds (see Figure 1). These scientific and academic developments were paralleled by exponential growth in the output of plastics, as shown by the graph in Figure 2. In short, the United States dominated polymer science and the polymer-based industries for thirty years after World War II.[1]

It is therefore not altogether surprising that the rubber research program identified itself with the development of polymer science from an early stage. The name of the discussion group was quickly changed from the *Rubber* Research Discussion Group to the *Polymer* Research Discussion Group. When the research section was transferred to the Rubber Reserve Company in 1944, it was renamed the *polymer* research section.

The supporters of the research program naturally stress the contribution made by the program to the development of polymer science. Maurice Morton has drawn attention to five "spin-offs," contributions

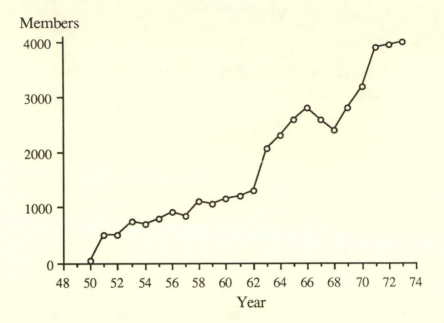

FIGURE 1. Growth of polymer division memberships.

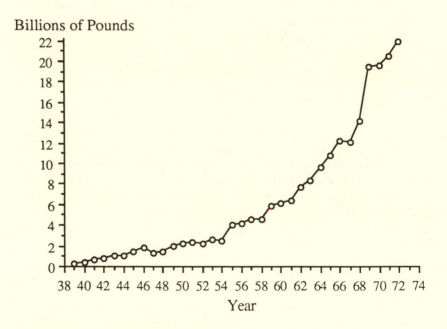

FIGURE 2. Resin and plastics production.

made by the program which "led to the rapid advances of polymer science that we have witnessed during the past 30 years."[2] All large, expensive scientific or technical projects like to claim that they generate "spin-offs," which thereby justify the huge expense involved. The best known example is the space program;[3] others include the ill-fated Mohole Project,[4] and the latest plans for a "supercollider" particle accelerator.[5] The rubber program is unusual in that its "spin-offs" were scientific by-products of an industrial development program rather than commercial "spin-offs" from a scientific project.

Nevertheless, we must be wary of falling into the common fallacy of *post hoc, ergo propter hoc*. The fact that polymer science grew rapidly in the postwar period (it actually *started* to grow much earlier) does not necessarily mean that the rubber program was the sole cause of this growth, or even the most important cause, but perhaps only one cause among several. In this chapter we will examine the "spin-offs" indicated by Morton[6] and how the program promoted the research of Paul Flory, the only U.S. polymer scientist to win the Nobel Prize. We then consider the program's contribution to the increase in the number of polymer scientists and the development of polymer education.

The Spin-Offs

Peter Debye's development of light scattering as a technique for determining high molecular weights was the rubber research program's most important contribution to polymer science. There were foreshadowings of Debye's theory in earlier papers on light scattering by solutions.[7] It is clear, however, that Debye's work in 1943 laid the foundations of the modern technique of light scattering.

Herbert Morawetz, an eminent historian of polymer science, has noted that "infra-red spectroscopy proved to be a particularly powerful tool for the study of polymers . . . it may be argued that its importance to polymer science was even greater than to the organic chemistry" of small molecules.[8] The rubber *production* program played a vital role in the adoption of this "powerful tool" through its assistance with the development of commercial spectrophotometers, especially the Beckman IR–1 models. The close cooperation between Shell Development Company and National Technical Instruments (and the feedback from other users at the 1943 summer school) produced a state-of-the-art instrument, which met the demands placed on it by industry but was also easy to use. Rabkin, in his recent paper on the development of infrared spectroscopy, remarked: "The cooperative nature of the wartime infrared work that characterized the U.S. rubber program . . . resulted in fast and effective diffusion of infrared spectroscopy."[9]

The clarification of the copolymerization of butadiene and styrene was obviously important for the rubber research program. Interest in this question was not limited to those working in the program, because of its contribution to the elucidation of polymerization in general and the industrial importance of other copolymers, for example copolymers of vinyl chloride and vinyl acetate. Frank Mayo and his research group at U.S. Rubber played a major role in the development of the theory of copolymerization. Mayo clearly benefitted from his access to the research program reports, but his own work was considered by U.S. Rubber to fall within the company's private research, and it was not sponsored by the government. Paul Flory was in a similar situation while he was working for Goodyear. The history of copolymerization theory provides an interesting example of a topic that was developed over a decade with alternating contributions from within and outwith the program.[10]

Even before the synthetic rubber program, Fred Wall of Illinois had suggested in 1941 that the absolute rates of reaction be ignored in favor of the ratio of the relative rates at which the two monomers react.[11] He went on to suggest that this ratio was proportional to the rate of addition of the two monomers. Marvel carried out experiments that supported Wall's simple relationship,[12] but it was disproved by Mayo.[13] Attention was now given to the development of a more sophisticated equation. The so-called copolymerization equation was presented in three independent papers published in 1944, by Turner Alfrey and George Goldfinger,[14] Mayo and Frederick Lewis,[15] and by Wall himself.[16] This equation related the composition of the copolymer at any given instant (the ratio suggested by Wall in 1941) to the rate of addition by using two parameters, which were called the "monomer reactivity ratios."

As the experimental data mounted, Mayo and his group concluded in 1945 that there was no connection between the rate at which a monomer polymerized on its own and its reactivity ratio.[17] Furthermore, there was a tendency for the two comonomers to alternate in the polymer. Mayo suggested that the explanation could be a combination of steric and electronic effects. This explanation was taken up by Charles Price, while he was a visiting professor at Brooklyn Polytechnic in 1946, and his new office-mate, Turner Alfrey. The Alfrey-Price equation related the reactivity ratios to the general reactivity and polar properties of the two monomers.[18]

There was not, as yet, any valid model for predicting the absolute rates. By reworking the assumptions behind the copolymerization equation, Mayo's colleague Cheves Walling developed a set of equations for the absolute rates of copolymerization in 1949.[19] He introduced a new parameter related to the termination processes. The British chemist (Sir)

Harry Melville at Aberdeen University had published a similar equation for the absolute rate of copolymerization in 1947.[20]

In 1950, Mayo and his group dramatically demonstrated the variation produced in copolymers by changes in the polymerization mechanism.[21] When an equimolar feed of styrene and methyl methacrylate undergoes free radical polymerization (using a peroxide) a 1:1 copolymer is formed. By contrast, cationic polymerization with stannic chloride produces a nearly pure polystyrene, and anionic polymerization with sodium or potassium gives almost pure polymethyl methacrylate.

The elucidation of the mechanism of the emulsion polymerization of GR-S was another substantial scientific achievement of the rubber research program. William Draper Harkins finally received the international recognition he had been denied for his earlier research on surface tension and nuclear chemistry. The Harkins-Smith-Ewart (HSE) theory of emulsion polymerization dominated the field for nearly two decades, mainly because much of the subsequent experimentation was on systems similar to GR-S. It is interesting to note, however, that R. N. Haward, at the British firm Petrochemicals Ltd., published a similar treatment in 1949.[22]

Nevertheless, Harkins made a fundamental error in locating the starting point of the polymerization process in the small micelles. The HSE theory has been replaced by the more comprehensive and accurate homogeneous nucleation model of emulsion polymerization.[23] Hans Fikentscher of I.G. Farben had been the first to suggest, in 1938, that the initial polymerization took place in the aqueous phase.[24] This view was supported by the study of the kinetics of styrene polymerization published by Charles Price and his student Clark Adams in 1945.[25] The basic theory was proposed by M. G. Evans and J. H. Baxendale of Leeds University, England, in 1946, as a result of their research on methyl methacrylate.[26] It was elaborated in 1952 by William Priest, who was working on the polymerization of vinyl acetate at Eastman Kodak.[27] Priest's scheme, following a decade of eclipse by the HSE theory, was confirmed by several groups in the 1960s.[28] A quantitative theory based on Priest's model was produced in 1970 by Robert Fitch and C. H. Tsai at the University of Connecticut.[29]

The weaknesses of the HSE theory should not be allowed to obscure the contribution it made to the synthetic rubber program. As the first detailed model of emulsion polymerization, this theory was a major scientific achievement when it was first published. Indeed, as it is relatively accurate for the special case of styrene-butadiene emulsion polymerization, it is still used in the rubber industry.

The record of the rubber research program is generally positive. Nonetheless, there were several missed opportunities. Given the importance of gel for the properties of GR-S, it might be thought that the theory of gel formation would be an important aspect of the rubber research. The basic theory of gelation had been laid out by Paul Flory in 1941.[30] However, his theory was concerned with polycondensation, because it was based on his work with polyesters at Du Pont rather than on the free radical polymerization of dienes.

In 1943, Walter Stockmayer, working with Joseph Mayer at Columbia, independently derived a similar theory from a parallel with the condensation of gases.[31] Stockmayer's work owed more to Mayer's interest in the applications of statistical mechanics than contemporary developments in polymer science. In 1944, Stockmayer extended his theory to the cross-linking of polydienes.[32] This was elaborated by Flory in 1947, when he published a paper on gel formation by cross-links in diene polymers.[33] Flory's former colleagues at Goodyear admit that they did not use his theory nor did they seek to involve him in the solution of the gel problem.[34] However, Maurice Morton, a newcomer at the University of Akron, applied Flory's treatment of diene gelation to GR-S in 1948.[35] While this work gave a scientific underpinning to the empirical research of the rubber companies, it had no impact on the program since cold rubber was already in production.

In keeping with the program's practical mission, its major contribution in this field was a better understanding of the effect of gel, especially microgel, on the properties of the final product. Only a few years after total confusion had existed about the nature and formation of gel in rubber, the American rubber program was able to produce a gel-free rubber. Baker's research at BTL had shown that "the condensed phase processibility, visco-elastic properties and eventual vulcanizate quality of rubber are dramatically dependent on the presence and quality of the microgel molecules." As a result of the research at BTL and U.S. Rubber, it was even possible to make a rubber with a predetermined proportion of gel (GR-S 60), which was better for extruded goods than normal GR-S. Several years later, the British Rubber Producers' Research Association created a natural rubber microgel by chemical treatment of raw latex for the same purpose.[36]

We might have expected the rubber research program to have materially assisted the development of rheology—the study of the deformation and stressing of materials—and the theory of rubber elasticity. However, James L. White comments in a recent essay on rheology:

Rheology played a relatively minor role in this program but that role was dominated in large part by [Melvin] Mooney [of U.S. Rubber] utilizing and expanding the concepts devised during the previous decade. Mooney's 1934 shearing disk viscometer was established by the Rubber Reserve as a quality control instrument, which it remains to this day. Mooney investigated a wide range of problems. . . . In succeeding years little of this research was followed up by the rubber companies . . . and the interaction between polymer rheology and processing was developed in large part by plastics companies. . . . In 1948, Alfrey published a monograph on the mechanical behavior of polymers which had a major influence in succeeding years.[37]

A similar situation existed in the case of rubber elasticity theory. Herman Mark and Eugene Guth, at the University of Vienna, had published the first account of the statistical mechanical or kinetic theory of rubber elasticity in 1934.[38] A comparable theory was published by Werner Kuhn in 1936.[39] A more sophisticated network theory was presented at an ACS meeting by Guth (who had moved to Notre Dame University) and Hubert James in 1939.[40] Guth and James continued to develop their theory over the next decade. The rubber program initially supported Guth's research but allowed his contract to lapse after the first year because the Rubber Research Board considered his work too theoretical.[41] Guth was also active (and possibly more influential) as a consultant to both Goodyear and Goodrich.[42]

Paul Flory studied the properties of the rubber network with his colleague John Rehner while he was still at Jersey Standard, and they published the Flory-Rehner theory in 1943.[43] They showed how the swelling of a vulcanized rubber (unvulcanized rubber lacks the cross-links necessary to form a network) by a solvent could be used to estimate the amount of cross-linking. In the same year, Flory left for Goodyear, where he continued to work on the relationship between the cross-link density of the network and the physical properties of the rubber. This research was supported, in part, by a grant from the Office of Naval Research. In 1951, Flory collaborated with Fred Wall to produce a paper on the statistical thermodynamics of rubber elasticity.[44]

An almost unique combination of chemical expertise, manpower, and facilities permitted the rubber research program to assist the rapid advance of polymer science in the 1940s and early 1950s. Nevertheless, it was only one component of a much larger international process. To give a final example, a small group of British chemists made great strides with the development of free radical polymerization in the same period. Harry Melville studied the kinetics of free radical polymerization in the 1940s, as an extension of his earlier work on free radical reactions in the gas phase.[45] He used the rotating sector technique, in contrast to Kolthoff's polarographic method. Melville was able to measure the absolute propagation and termination rates, which was a major step forward.[46] At a Faraday Society meeting on "Oxidation" in 1945, Reginald Bacon presented the results of five years' research on redox polymerization at ICI's Blackley laboratory.[47] The Blackley group collaborated with M. G. Evans, who gave a report on the polymerization of vinyl compounds with ferrous sulfate and hydrogen peroxide.[48] Another major figure in the British school of free radical polymerization studies was Clement Bamford—leader of Courtaulds's long-term research group based in Maidenhead—who discovered long-lasting free radicals, which were trapped in polyacrylonitrile.[49]

Flory and the Rubber Research Program

Paul Flory was one of the truly great scientists of the twentieth century.[50] We have already noted his research on gelation and rubber elasticity. In this period, he was also a pioneer in the theory of polymer configurations, polymer crystallization, liquid crystals, and even light scattering. Flory worked for three institutions (Jersey Standard, Goodyear, and Cornell University) that participated in the rubber research program. He regularly attended the meetings of the Polymer Research Discussion Group. In 1943, he was chairman of the Temporary Committee on Standardized Research Methods and a member of the Sub-Committee on Polymer Structure.

The contract between the Office of Rubber Reserve and Cornell made it possible for the university to give Flory the facilities he needed.[51] Flory himself received grants from the rubber program while he was on the Cornell faculty and carrying out some of his most important research, for example, his studies of the interaction between the polymer molecule and solvents with Thomas G. Fox.[52] For four years (1950–1953) these grants were a major part of his total external financial support.[53] In contrast to its usual policy of only funding research related to synthetic rubber in some way, the Office of Rubber Reserve apparently permitted Flory to use his funds for a variety of different, non-rubber–related topics. These included the study of heats of fusion of polyamides and polyesters,[54] the determination of the molecular dimensions of silicones,[55] and the intrinsic viscosity of polyelectrolytes.[56] This relationship was of great value to both sides, and to polymer science as a whole.

Nevertheless, it is clear that Flory would have still carried out his pathbreaking research in the absence of the rubber research program. His interest in butyl rubber, the network theory, and rubber physics, generally, was a result of his association with Jersey Standard before the rubber program began. Flory produced only a small number of CR reports during the wartime program, and he was not in the section of Goodyear research that was sponsored by the program. As far as the company was concerned, he was mostly working on the development of polyesters, a purely "private" area of research.[57] Flory also received a significant amount of funding from the Office of Naval Research and the Allegheny Ballistics Laboratory (operated by Hercules for the U.S. Navy).[58]

Polymer Education

The research program contributed to the development of polymer science indirectly by training graduate students as polymer scientists.[59] This was particularly true of the University of Illinois, which produced about a hundred program-related Ph.D.'s between 1945 and 1956, but not all these Ph.D.'s were in polymer science.[60] Furthermore, the annual total of Ph.D.'s for the whole program was usually no more than about a dozen or so, rising to perhaps thirty at its peak around 1950.

The earliest, and the preeminent, center of polymer education was Brooklyn Polytechnic. This institution (now called Polytechnic University, Brooklyn) has 146 surviving Ph.D.'s in polymer chemistry from the 1948–1956 period, which—allowing for attrition over the last thirty years—means that it produced almost twice as many Ph.D.'s as Illinois in polymer science–related topics in the early 1950s.[61] The polymer research group at Brooklyn, headed by Herman Mark, was never part of the rubber program. Raymond Seymour, another early pioneer of polymer education who taught polymer science at the University of Chattanooga and then at the University of Houston, also never took part in the program.

Nevertheless, the other two major centers in the late 1940s and early 1950s, Illinois and Cornell, were at the heart of the program. At least two eminent polymer chemists (William Bailey and William Krigbaum) received their Ph.D.'s working in the program at Illinois.

The academic community of polymer scientists in the 1950s contained several leading teachers who had been connected with the research program, but had not become fully fledged members. They include Charles Price (University of Pennsylvania), Arthur Tobolsky (Princeton), Eugene Guth (Notre Dame), Walter Stockmayer (MIT), and Leo Mandelkern (Florida State). Furthermore, there was a significant number of educators who had no contact with the program. John Ferry (Wisconsin), Raymond Fuoss (Yale), Malcolm Dole (Northwestern), George Butler (University of Florida), and Richard Stein (University of Massachusetts) were members of this group. Lehigh University was another early center of polymer education without any links to the program.

The crucial point is that the institutions in the program failed to establish themselves as the only key centers in polymer education or polymer research, or even to become the leaders, with the exception of Cornell during Flory's tenure. Furthermore, when an institution in the program did play a key role in the 1950s, it was sometimes due to an educator who was not associated with the earlier program-directed research there, such as Flory at Cornell or Stockmayer at MIT.

One major reason for this muted impact was the lack of a strategy by the central administration for the training of postgraduates within the program. At the beginning of the program, when the problems were both urgent and great, a very low priority was given to personnel training. This was no longer the case after the war, and in any event, the training of postgraduates would have not used scarce manpower in the way that training plant personnel did. The Illinois group produced at least six Ph.D.'s a year without undue strain on their resources.

At a time when polymer scientists were few in number and polymer education still in its infancy, the rubber program could have given a major impetus to the national development of the discipline at a relatively low cost by paying more attention to the training of young chemists. The National Science Foundation operated a very successful program in the mid-1960s when there was concern about the low number of well-qualified scientists in the United States.[62] It is surprising that the research program had such a limited impact, especially when the extra funds generated by the "overhead" element of the grants are taken into account.

The Polymer Research Discussion Group Revisited

The urgency and importance of the wartime research created a feeling of solidarity between the leading researchers in the rubber research program. This esprit de corps was maintained by the quarterly meetings of the Polymer Research Discussion Group (PRDG) and by correspondence between individual researchers. For example, William O. Baker corresponded with Piet Kolthoff and (occasionally) Paul Flory in the early years of the research program.[63] The friendships formed in this period survive to the present day. The veterans of the rubber research program continued to meet at meetings of the Polymer Division and Rubber Division of the American Chemical Society and the Gibson Island (later Gordon Research) Conferences. The postwar stresses that shattered the community of scientists created by the Manhattan project—policy disputes over the H-bomb and contrasting reactions to the dropping of the A-bomb—were not mirrored in the rubber program. Indeed, the success of the wartime production program, and the continuation of the research program until 1956, increased the group's solidarity.

While the PRDG doubtlessly assisted the postwar development of polymer science, its influence should not be exaggerated. The ACS Rubber Division, the High Polymer Forum (founded by Calvin Fuller and Adolf Elm, its first meeting held in 1946[64]), and its successor, the ACS Polymer Division, and the Gordon Research Conferences played major roles in the development of polymer science. During World War II, Herman Mark established his celebrated Saturday morning seminars at Brooklyn Polytechnic, which served as a bridge between those working in the rubber research program and other polymer scientists.[65]

If the PRDG aided cross-fertilization, we might expect to find some evidence of this in joint papers by coauthors who had no preprogram associations, either during or after the program. In the early 1950s, the staff of the Reconstruction Finance Corporation (RFC) compiled abstracts of scientific publications between 1942 and 1953, which stemmed from research funded by the rubber program.[66] The abstracts of the seven hundred or so papers thus recorded were examined for joint papers. Communications from the RFC, usually short communiqués, and papers that had only an employee of the RFC as an "external" coauthor were discarded.

A total of twenty-four papers with coauthors from different institutions, not linked by a student/professor or similar relationship, were found. However, thirteen of these papers were from the Government Laboratories at Akron. Of these thirteen, five were connected with new polymers prepared at the University of Illinois, three reported research carried out in collaboration with Goodrich, and another three stemmed from work with the laboratories of the U.S. Department of Agriculture. Seven of the remaining eleven papers were a result of scientific cooperation between different research groups. This low figure is confirmed by a similar examination of the publications (from 1942 onwards) of four leading scientists in the program: Flory, Kharasch, Kolthoff, and Marvel.[67] This survey yielded only one paper, the 1951 paper by Wall and Flory mentioned earlier.

A parallel situation is found if the careers of the participants in the research program are examined.[68] A detailed study of 152 younger scientists (who received their Ph.D.'s between 1934 and 1948, and who worked in the program between 1943 and 1946) revealed only one apparent pattern. Archie Deutschman, who worked briefly with Marvel during the war, became a professor of biochemistry at the University of Arizona in 1957. Myron Corrin, a former student of Harkins, was appointed professor of chemistry in 1959. Two years later, Carl Marvel was made a research professor there, when he retired at Illinois. We have already noted that Paul Flory joined the faculty of Cornell University in 1948, largely as result of the facilities made possible by the rubber research program.

It is not altogether surprising that the PRDG made only a modest contribution to the development of polymer science. The PRDG was created during the wartime emergency to coordinate the work of the various groups and to allow the researchers to know each other better. The agenda was therefore limited (despite the group's title) to matters of interest to the rubber program. The Polymer Research Discussion Group was organized for the benefit of the rubber program, rather than the advancement of polymer science.[69]

A Missed Opportunity?

The rubber research program made many noteworthy contributions to polymer science. It trained a significant proportion of graduate students in this field. The research program promoted a sense of solidarity between the researchers and thereby assisted the development of polymer science as a discipline. It had a enduring influence on the research and development programs of the rubber companies, which were henceforth associated with polymer science, rather than the empirical methods of the prewar industry. It also demonstrated the value of polymer science to the Armed Forces. The Office of Naval Research and the U.S. Air Force soon became major sponsors of polymer research, largely as a result of the rubber program.

Nonetheless, a degree of caution is required. Similar contributions were made, in the same areas of polymer science, elsewhere around the same time. The research program did not break into completely new areas, with the possible exceptions of molecular networks (microgel) and light scattering. Furthermore, the program was weak in areas in which it would be expected to be strong, for example, rheology, elasticity, and gelation theory.

A similar conclusion can be reached for polymer education. Several of the key figures in the development of polymer education in the postwar period were not connected with the research program. Only two of the institutions funded by the program (Illinois and Cornell) were important in this respect, and they were more than counterbalanced by other colleges external to the program: Brooklyn Polytechnic, the University of Florida, Yale University, and Lehigh University. The relative failure of the program-associated universities can be traced, in large measure, to the lack of a postwar strategy for the education of younger chemists.

No program should be blamed for failing to achieve objectives over and above those it was created to fulfill. The rubber research program was established to overcome the problems associated with the original GR-S and to develop a better synthetic rubber. It was not set up to promote polymer science, in the sense that the National Institutes of Health were established to advance the biomedical sciences. Even so, the 1940s and 1950s were a crucial period in the development of polymer science. The rubber program, imperfect as it was, contributed to America's success in this field. If the program's wartime mission had been replaced by a strategy for the development of polymer science as a whole, its impact might have been greater and longer lasting. Instead of being highlighted alongside atomic energy as a national concern, of importance to national security and economic growth, polymer science was subsumed into the National Science Foundation.

Notes

1. The graphs in Figures 1 and 2 are reproduced from R. D. Ulrich, "History of the ACS Division of Polymer Chemistry, Inc.," in Ulrich (ed.), *Contemporary Topics in Polymer Science,* volume 1, *Macromolecular Science: Retrospect and Prospect* (New York and London, 1978), 3.

For surveys of the history of polymer science, also see: Herman Mark, "Polymers—Past, Present, Future," in W. O. Milligan (ed.), *Proceedings of The Robert A. Welch Foundation Conference on Chemical Research. X: Polymers* (Houston, Texas, 1967), 19–55; Mark, "Polymer Chemistry: The Past Hundred Years," ACS Centennial Issue *Chemical and Engineering News* 54 (6 April 1976), 176–189; Mark, "Polymer Chemistry in Europe and America—How It All Began," *Journal of Chemical Education* 58 (1981), 527–534. C. S. Marvel, "The Development of Polymer Chemistry in America: The Early Days," *Journal of Chemical Education* 58 (1981), 535–539. Herbert Morawetz, *Polymers: The Origins and Growth of a Science* (New York, 1985). Raymond B. Seymour (ed.), *History of Polymer Science and Technology* (New York and Basel, 1982). G. Allan Stahl (ed.), *Polymer Science Overview: A Tribute to Herman F. Mark,* ACS Symposium Series 175 (Washington, D.C., 1981); Stahl, "Development of Modern Polymer Theory," *CHEMTECH* 14 (1984), 492–495. Walter H. Stockmayer and Bruno H. Zimm, "When Polymer Science Looked Easy," *Annual Review of Physical Chemistry* 35 (1984), 1–21.

Also see the interviews of Herman Mark by James Bohning and Jeffrey Sturchio for the Beckman Center oral history program, 3 February 1986, 17 March 1986, and 20 June 1986.

2. M. Morton, "History of Synthetic Rubber," 236.

3. NASA even publishes an annual volume entitled *Spinoff.*

4. Milton Lomask, *A Minor Miracle: An Informal History of the National Science Foundation* (Washington, D.C., 1976), 167–168.

5. Sheldon Glashow and Leon Lederman, "The SSC: A Machine for the Nineties," *Physics Today* 38 (March 1985), 29–37.

6. M. Morton, "History of Synthetic Rubber," 236.

7. Morawetz, *Polymers,* 111. M. Fixman, "Molecular Theory of Light Scattering," *Journal of Chemical Physics* 23 (1955), 2074–2079. B. H. Zimm, *Journal of Chemical Physics* 13 (1945), 141–145. B. Zimm, R. S. Stein, and P. Doty, "Classical Theory of Light Scattering from Solutions—A Review," *Polymer Bulletin* 1 (1945), 90–119. These papers were reprinted in D. McIntyre and F. Gornick (eds.), *Light Scattering from Dilute Polymer Solutions* (New York, 1964). Also see the interview of Bruno Zimm by James Bohning for the Beckman Center oral history program, 9 September 1986.

8. Morawetz, *Polymers,* 214–215.

9. Rabkin, "Technological Innovation in Science," 47.

10. Frank R. Mayo and Cheves Walling, "Copolymerization," *Chemical Reviews* 46 (1950), 190–287, is an excellent account of the copolymerization research up to 1950.

11. F. T. Wall, *Journal of the American Chemical Society* 63 (1941), 1862–1866.

12. C. S. Marvel, G. D. Jones, T. W. Mastin, and G. L. Schertz, *Journal of the American Chemical Society* 64 (1942), 2356–2362; Marvel and Schertz, *Journal of the American Chemical Society* 65 (1943), 2054–2058; Marvel and Schertz, *Journal of the American Chemical Society* 66 (1944), 2135.

13. Mayo and Walling, "Copolymerization," 195.

14. T. Alfrey, Jr., and G. Goldfinger, *Journal of Chemical Physics* 12 (1944), 205–209.

15. F. R. Mayo and F. M. Lewis, *Journal of the American Chemical Society* 66 (1944), 1594–1601.

16. F. T. Wall, *Journal of the American Chemical Society* 66 (1944), 2050–2057.

17. F. M. Lewis, F. R. Mayo, and W. F. Hulse, *Journal of the American Chemical Society* 67 (1945), 1701–1705.

18. T. Alfrey, Jr., and C. C. Price, *Journal of Polymer Science* 2 (1947), 101–106. Additional information from Charles C. Price, 28 May 1987.

19. C. J. Walling, *Journal of the American Chemical Society* 71 (1949), 1930–1935.

20. H. W. Melville, B. Noble, and W. F. Watson, *Journal of Polymer Science* 2 (1947), 229–245.

21. C. J. Walling, E. R. Briggs, W. Cummings, and F. R. Mayo, *Journal of the American Chemical Society* 72 (1950), 48–51.

22. R. N. Haward, *Journal of Polymer Science* 4 (1949), 273–287. Haward was a former student of M. G. Evans at the University of Leeds.

23. R. M. Fitch, "Latex Particle Nulceation and Growth," in D. R. Bassett and A. E. Hamielec (eds.), *Emulsion Polymers and Emulsion Polymerization*, ACS Symposium Series 165 (Washington, D.C., 1981), 1–6.

24. H. Fikentscher, *Angewandte Chemie* 51 (1938), 433.

25. C. C. Price and C. E. Adams, *Journal of the American Chemical Society* 67 (1945), 1674–1680.

26. J. H. Baxendale, M. G. Evans, and J. K. Kilham, *Transactions of the Faraday Society* 42 (1946), 668–675. Baxendale, S. Bywater, and Evans, *Transactions of the Faraday Society* 42 (1946), 675–684.

27. W. J. Priest, *Journal of Physical Chemistry* 56 (1952), 1077–1082.

28. For example, by A. S. Dunn and P. A. Taylor, *Makromolekulare Chemie* 83 (1965), 207–210.

29. R. M. Fitch and C. H. Tsai, *Journal of Polymer Science: B. Polymer Letters* 8 (1970), 703–710.

30. P. J. Flory, *Journal of the American Chemical Society* 63 (1941), 3083–3090. For a commentary on the relationship between gelation theory and the rubber program, see M. Morton, "Rubber Enters the Polymer Age," *Rubber Chemistry and Technology* 58 (1985), G78–G80.

31. W. H. Stockmayer, *Journal of Chemical Physics* 11 (1943), 45–55. Interview of Walter Stockmayer by Peter Morris and Jeffrey Sturchio for the Beckman Center oral history program, 25 August 1986.

32. W. H. Stockmayer, *Journal of Chemical Physics* 12 (1944), 125–131.

33. P. J. Flory, *Journal of the American Chemical Society* 69 (1947), 30–35.

34. Gehman interview and Pierson interview.

35. Morton, "Rubber Enters the Polymer Age," G78.

36. W. O. Baker, "Microgel Molecules"; quotation on 1–2. Also see D. L. Schoene, A. J. Green, E. R. Burns, and G. R. Vila, *Industrial and Engineering Chemistry* 38 (1946), 1246–1249.

37. J. L. White, "Rheological Behavior of Unvulcanized Rubber," Chapter 6 of F. R. Eirich (ed.), *Science and Technology of Rubber* (New York, 1978), 224–225.

38. E. Guth and H. Mark, *Monatshefte Chemie* 65 (1934), 93–121. Also see E. Guth "Birth and Rise of Polymer Science—Myth and Truth," *Journal of Applied Polymer Science: Applied Polymer Symposium* 35 (1979), 1–12.

39. W. Kuhn, *Angewandte Chemie* 49 (1936), 858–862.

40. E. Guth and H. M. James, *Industrial and Engineering Chemistry* 33 (1941), 624–629. Later papers included: H. M. James and E. Guth, *Journal of Chemical Physics* 11 (1943), 455–481; James and Guth, *Journal of Chemical Physics* 15 (1947), 669–683; James and Guth, *Journal of Polymer Science* 4 (1949), 153–182; and James and Guth, *Journal of Chemical Physics* 21 (1953), 1039–1049.

41. "Summary . . . of Meeting No. 9 held September 9, 1943," RG 234, RRC, Entry 231, PI–173, Minutes of the Rubber Research Board, 5–6.

42. Gehman interview and Beatty interview.

43. P. J. Flory and J. Rehner, Jr., *Journal of Chemical Physics* 11 (1943), 512–526. Also see Flory, "Network Structure and the Elastic Properties of Vulcanized Rubber," *Chemical Reviews* 35 (1944), 51–75.

44. F. T. Wall and P. J. Flory, *Journal of Chemical Physics* 19 (1951), 1435–1439.

45. His first study of polymerization was the gas-phase polymerization of methyl methacrylate. H. W. Melville, *Proceedings of the Royal Society: A* 163 (1937), 511–542.

46. Morawetz, *Polymers,* 172. G. M. Burnett and H. W. Melville, *Nature* 156 (1945), 661; Burnett and Melville, *Proceedings of the Royal Society: A* 189 (1947), 456–480.

47. R. G. R. Bacon, *Transactions of the Faraday Society* 42 (1946), 140–155.

48. J. H. Baxendale, M. G. Evans, and G. S. Parks, *Transactions of the Faraday Society* 42 (1946), 155–169.

49. C. H. Bamford and A. D. Jenkins, *Proceedings of the Royal Society: A* 216 (1953), 515–539; Bamford and Jenkins, *Proceedings of the Royal Society: A* 228 (1955), 220–237. For a generous appraisal of the polypeptide research at Maidenhead, see Morawetz, *Polymers,* 195.

50. "Paul John Flory" in Morris, *Polymer Pioneers,* 70–73. Also see Walter H. Stockmayer, "The 1974 Nobel Prize for Chemistry," *Science* 186 (1974), 724–726; and the "Autobiographical Note" at the beginning of Leo Mandelkern, James E. Mark, Ulrich W. Suter, and Do Y. Yoon, *Selected Works of Paul J. Flory,* volume 1 (Stanford, California, 1985).

51. Private communication from Leo Mandelkern, 11 August 1987.

52. P. J. Flory and T. G. Fox, Jr., *Journal of the American Chemical Society* 73 (1951), 1904–1908.

53. In the period between 1947 and 1958 (external grants were not significant before 1947) the financial assistance of the rubber research program was acknowledged in twenty-three of a total of of seventy-one papers published in L. Mandelkern, et al., *Selected Works of Paul J. Flory,* three volumes.

54. P. J. Flory, L. Mandelkern, and H. K. Hall, *Journal of the American Chemical Society* 73 (1951), 2532–2538. Mandelkern, R. R. Garrett, and Flory, *Journal of the American Chemical Society* 74 (1952), 3949–3951.

55. P. J. Flory, L. Mandelkern, J. B. Kinsinger, and W. B. Schultz, *Journal of the American Chemical Society* 74 (1952), 3364–3367.

56. P. J. Flory and J. E. Osterheld, *Journal of Physical Chemistry* 58 (1954), 653–661.

57. Pierson interview.

58. The support of the Office of Naval Research was acknowledged in eleven papers (ten arising from a contract between the ONR and Goodyear) in Mandelkern, et al., *Selected Works of Paul J. Flory,* between 1947 and 1958, and the Allegheny Ballistics Laboratory in thirteen more.

59. C E. Carraher, Jr., "Polymer Education and the Mark Connection," in Stahl (ed.), *Polymer Science Overview,* 123–142.

60. Calculated from a computer print-out of University of Illinois chemistry Ph.D.'s, 1943–1956, obtained from Comprehensive Dissertation Index Online.

61. Computer print-out of Brooklyn Polytechnic alumni in polymer chemistry provided by the university for the Beckman Center oral history program.

62. These graduate trainee fellowships, started in 1964, were a result of a 1962 report from a panel of the President's Science Advisory Committee headed by Edwin Gilliland, who had been the leader of the rubber research and development program in 1944. The fellowships were originally instituted for engineering, but extended to all subject areas covered by the NSF in 1966. Lomask, *A Minor Miracle,* 130.

63. See correspondence between 1943 and 1947 in casefile 24060–1, AT&T Bell Laboratories Archives.

64. Carraher, "Polymer Education," 133. Information from C. S. Fuller, 24 July 1987.

65. G. Allan Stahl, "Herman F. Mark: The Geheimrat," in Stahl (ed.), *Polymer Science Overview,* 82.

66. "Abstracts of Technical Papers From the Government Synthetic Rubber Program," four volumes, Reconstruction Finance Corporation, Office of Rubber Reserve, 31 December 1953, RG 234, RRC, Entry 259, PI–173.

67. The bibliographies were supplied by the chemists concerned and are stored in the appropriate oral history files at the Beckman Center for History of Chemistry, with the exception of Morris Kharasch's bibliography, which was taken from Waters (ed.), *Vistas in Free Radical Chemistry,* 8–16.

68. This information was obtained from the prosopographical survey referenced in Chapter 3, note 127.

69. This paragraph is based, in part, on an argument put forward by C. S. Fuller in a private communication to Peter Morris, 24 July 1987.

Summary of Major Conclusions

1. The wartime rubber program was a well-directed emergency program which greatly improved the original GR-S and produced it on a large scale. By contrast, the broader postwar research program—to quote Herbert and Bisio—"achieved only mundane results."[1]

2. The academics associated with the synthetic rubber program were eager to help the rubber industry, but the rubber companies were slow to respond and did not always collaborate whole-heartedly. Nevertheless, the university-based researchers made a valuable contribution to the development of GR-S and cold rubber, especially in the related fields of chemical analysis, kinetics, and mechanistic studies. The introduction of light scattering as a method of molecular weight determination was a major step forward, and the heat-resistant fiber PBI stemmed from Marvel's search for new synthetic rubbers.

3. After the end of the research program in the mid-1950s, most of the academic scientists collaborated with the military research branches and/or chemical companies, rather than the rubber companies. Hardly any of the academics or graduate students who worked in the research program later pursued careers in the rubber industry.

4. Once World War II was over, the researchers published over seven hundred papers and thereby made significant contributions to the development of polymer science in the late 1940s and 1950s, "the golden age of polymer chemistry," as Cheves Walling recently remarked.[2] The key institution in the 1940s, however, was the Polymer Research Institute at Brooklyn Polytechnic, under the leadership of Herman Mark. Of the many institutions funded by the postwar research program, only Cornell and Illinois were of the first rank in the field of polymer science.

5. These conclusions are consonant with the hypothesis that government-sponsored programs involving industry-wide cooperation promote fundamental research and incremental improvements, rather than radical breakthroughs. This hypothesis does not reduce the importance of government funding of technology. Most, if not all, radical innovations are the unforeseen result of many years of fundamental research.

6. These conclusions are consonant with the further hypothesis that radical innovation is best promoted by competition. Competition can take different forms. It need not be commercial competition. Space-flight and military technology have been stimulated by superpower rivalry. Nonetheless, commercial competition remains the best means of accelerating technological innovation in a free society. This spur will be most effective when the necessary infrastructure already exists, namely several companies—above an approximate threshold size—with significant research and development programs. This precondition is met, for example, by the chemical, pharmaceutical, and electronics industries. It thus follows that mergers will often retard radical innovation by reducing competition.

Notes

1. Herbert and Bisio, *Synthetic Rubber,* 218.
2. Cheves Walling, "The Study of Polymers" (a review of Morawetz, *Polymers*), *Science* 231 (10 January 1986), 167.

Appendix: Introduction to Polymer Chemistry

In the realm of chemistry, many *compounds*—such as water or methane (natural gas)—are made up of a relatively small number of *atoms*. Water has only three atoms, two hydrogen atoms and one oxygen atom. Methane has five atoms, four hydrogen atoms and one carbon atom. We can draw their structures like this:

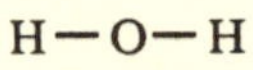

$$H-O-H$$

water (H_2O)

$$\begin{array}{c} H \\ | \\ H-C-H \\ | \\ H \end{array}$$

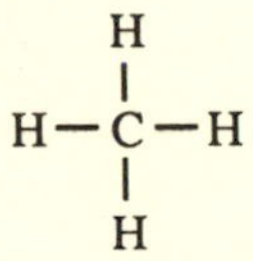

methane (CH_4)

The hydrogen atoms in methane can be replaced by other atoms or groups of atoms. For instance, if we replace three hydrogen atoms by chlorine atoms, we obtain the anesthetic chloroform. Similarly, if we insert an oxygen atom between the carbon and one of the hydrogens, we obtain a hybrid of water and methane that is called methanol or methyl alcohol (wood spirit):

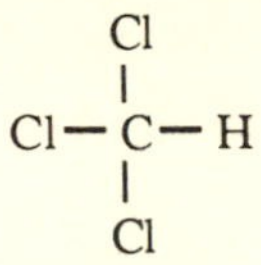

$$\begin{array}{c} Cl \\ | \\ Cl-C-H \\ | \\ Cl \end{array}$$

chloroform

$$\begin{array}{c} H \\ | \\ H-C-O-H \\ | \\ H \end{array}$$

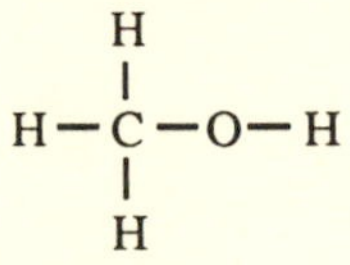

methanol

Carbon also has the unusual property of forming strong bonds with other carbon atoms. Hydrogen peroxide is unstable, decomposing readily to water and oxygen, whereas ethane is a stable gas found alongside methane in natural gas:

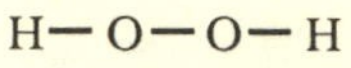

$$H-O-O-H$$

hydrogen peroxide

$$H-\overset{\displaystyle H}{\underset{\displaystyle H}{C}}-\overset{\displaystyle H}{\underset{\displaystyle H}{C}}-H$$

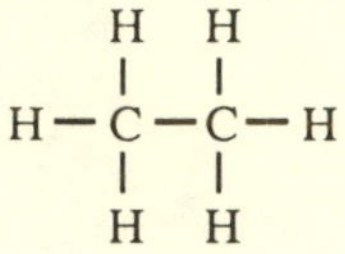

ethane

You will notice that carbon always has four bonds and oxygen two. We say that carbon has a *valency* of four, but oxygen has a valency of two. If we "fuse" ethane and water—as we did for methanol—we obtain ethanol or ethyl alcohol, the basis of alcoholic drinks like beer or whiskey:

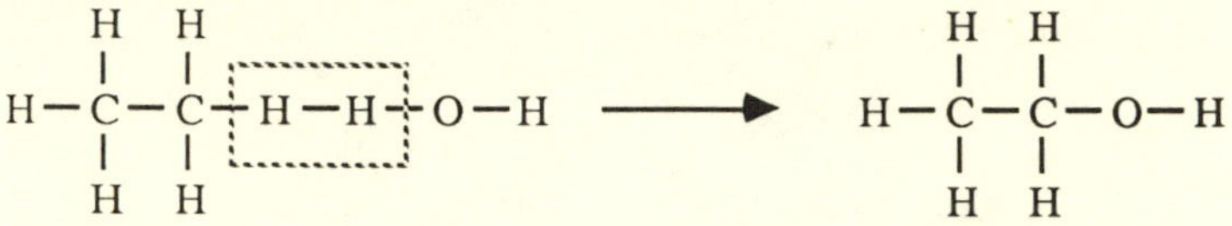

$$H-\overset{\displaystyle H}{\underset{\displaystyle H}{C}}-\overset{\displaystyle H}{\underset{\displaystyle H}{C}}\!-\!H-H\!-\!O-H \longrightarrow H-\overset{\displaystyle H}{\underset{\displaystyle H}{C}}-\overset{\displaystyle H}{\underset{\displaystyle H}{C}}-O-H$$

ethyl alcohol

However, ethyl alcohol is not prepared this way, but by adding water to another two-carbon compound, ethylene:

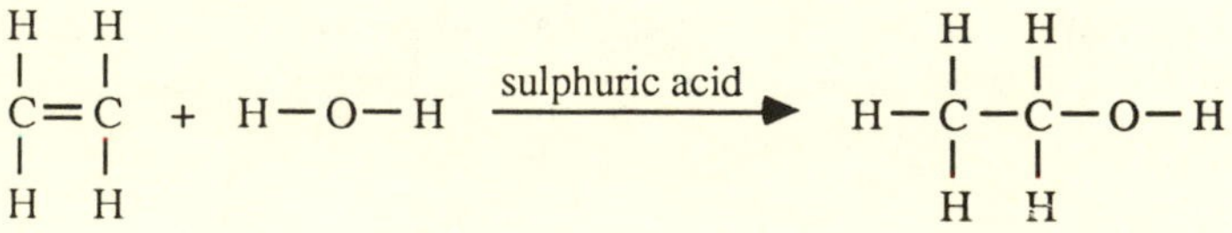

$$\overset{\displaystyle H}{\underset{\displaystyle H}{C}}=\overset{\displaystyle H}{\underset{\displaystyle H}{C}} \;+\; H-O-H \xrightarrow{\text{sulphuric acid}} H-\overset{\displaystyle H}{\underset{\displaystyle H}{C}}-\overset{\displaystyle H}{\underset{\displaystyle H}{C}}-O-H$$

ethylene   ethyl alcohol

Ethylene illustrates another unusual property of carbon, the ability to form two bonds with another carbon atom. This type of carbon-carbon bond is called a *double bond*. It is very difficult to break the single bond in ethane, but relatively easy to open the double bond in ethylene.

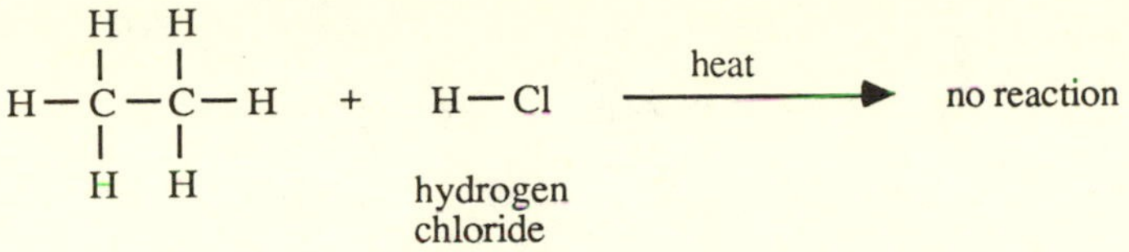

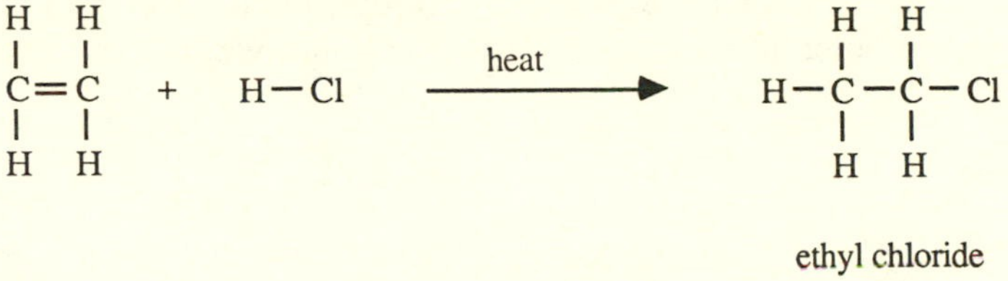

$$
\begin{array}{c}
\quad H \quad H \\
\quad | \quad\; | \\
H-C-C-H \\
\quad | \quad\; | \\
\quad H \quad H
\end{array}
\; + \;
\begin{array}{c}
H-Cl \\
\text{hydrogen} \\
\text{chloride}
\end{array}
\quad\xrightarrow{\;\text{heat}\;}\quad
\text{no reaction}
$$

$$
\begin{array}{c}
H \quad H \\
| \quad\; | \\
C = C \\
| \quad\; | \\
H \quad H
\end{array}
\; + \;
H-Cl
\quad\xrightarrow{\;\text{heat}\;}\quad
\begin{array}{c}
\quad H \quad H \\
\quad | \quad\; | \\
H-C-C-Cl \\
\quad | \quad\; | \\
\quad H \quad H
\end{array}
$$

ethyl chloride

It can be seen that ethyl chloride bears the same relationship to hydrogen chloride as ethyl alcohol to water. If you compress ethylene to a very high pressure (several thousand times more than atmospheric pressure) in the presence of a small trace of oxygen, a waxy solid is formed. This is the simplest *polymer*, polyethylene, which contains thousands of ethylene molecules joined together like beads on a very long string:

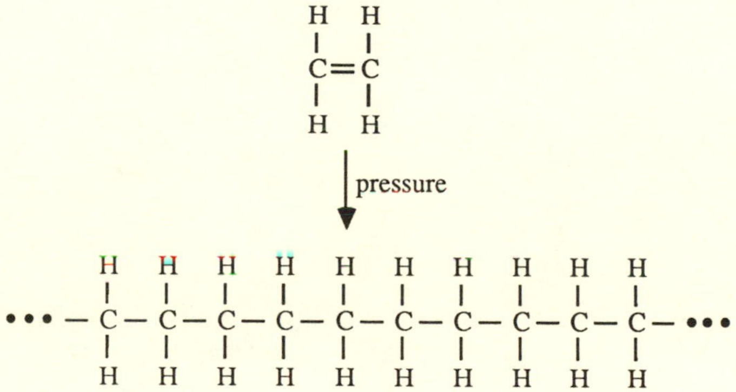

$$
\begin{array}{c}
H \quad H \\
| \quad\; | \\
C = C \\
| \quad\; | \\
H \quad H
\end{array}
$$

$$\big\downarrow\ \text{pressure}$$

$$
\cdots -
\begin{array}{c}
H \\ | \\ C \\ | \\ H
\end{array}
-
\begin{array}{c}
H \\ | \\ C \\ | \\ H
\end{array}
-
\begin{array}{c}
H \\ | \\ C \\ | \\ H
\end{array}
-
\begin{array}{c}
H \\ | \\ C \\ | \\ H
\end{array}
-
\begin{array}{c}
H \\ | \\ C \\ | \\ H
\end{array}
-
\begin{array}{c}
H \\ | \\ C \\ | \\ H
\end{array}
-
\begin{array}{c}
H \\ | \\ C \\ | \\ H
\end{array}
-
\begin{array}{c}
H \\ | \\ C \\ | \\ H
\end{array}
-
\begin{array}{c}
H \\ | \\ C \\ | \\ H
\end{array}
-
\begin{array}{c}
H \\ | \\ C \\ | \\ H
\end{array}
- \cdots
$$

Ethylene is the *monomer* of polyethylene and the process of creating polyethylene from its monomer is called *polymerization*. Most, but not all, polymers are long strings of one or two monomers.

Hydrocarbons are compounds that contain only carbon and hydrogen. Ethane and ethylene are *hydrocarbons*, as well as methane:

$$H-\underset{\underset{H}{|}}{\overset{\overset{H}{|}}{C}}-H \qquad H-\underset{\underset{H}{|}}{\overset{\overset{H}{|}}{C}}-\underset{\underset{H}{|}}{\overset{\overset{H}{|}}{C}}-H \qquad H-\underset{\underset{H}{|}}{\overset{\overset{H}{|}}{C}}-\underset{\underset{H}{|}}{\overset{\overset{H}{|}}{C}}-\underset{\underset{H}{|}}{\overset{\overset{H}{|}}{C}}-\underset{\underset{H}{|}}{\overset{\overset{H}{|}}{C}}-H$$

methane   ethane   butane

Another hydrocarbon, with four carbon atoms, is butane, which is sold in canisters under pressure as cigarette lighter fuel. If we replace one single bond in butane with a double bond, we obtain butylene:

$$H-\underset{\underset{H}{|}}{\overset{\overset{H}{4|}}{C}}-\underset{\underset{H}{|}}{\overset{\overset{H}{3|}}{C}}-\underset{\overset{H}{2|}}{C}=\underset{\underset{H}{}}{\overset{\overset{H}{1|}}{C}} \qquad\qquad \underset{\underset{H}{|}}{\overset{\overset{H}{4|}}{C}}=\underset{\underset{H}{|}}{\overset{\overset{H}{3|}}{C}}-\underset{\overset{H}{2|}}{C}=\underset{\underset{H}{}}{\overset{\overset{H}{1|}}{C}}$$

butylene   butadiene

Now, if we also form a double bond between carbon atoms three and four, we obtain a compound with two double bonds, butadiene. In 1910, the English chemist F. E. Matthews discovered that sodium metal, which is very reactive, triggered the polymerization of butadiene to polybutadiene, which is a rubbery polymer:

$$\underset{\underset{H}{|}}{\overset{\overset{H}{|}}{C}}=\underset{\underset{}{|}}{\overset{\overset{H}{|}}{C}}-\underset{}{\overset{\overset{H}{|}}{C}}=\underset{\underset{H}{|}}{\overset{\overset{H}{|}}{C}}$$

butadiene

| sodium
| metal
↓

$$\bullet\bullet\bullet-\underset{\underset{H}{|}}{\overset{\overset{H}{|}}{C}}-\underset{}{\overset{\overset{H}{|}}{C}}=\underset{}{\overset{\overset{H}{|}}{C}}-\underset{\underset{H}{|}}{\overset{\overset{H}{|}}{C}}-\underset{\underset{H}{|}}{\overset{\overset{H}{|}}{C}}-\underset{}{\overset{\overset{H}{|}}{C}}=\underset{}{\overset{\overset{H}{|}}{C}}-\underset{\underset{H}{|}}{\overset{\overset{H}{|}}{C}}-\bullet\bullet\bullet$$

polybutadiene

The monomer of natural rubber is called isoprene, which is very similar to butadiene. One of the hydrogen atoms attached to the second carbon atom of butadiene is replaced by one carbon atom and three hydrogen atoms, which is called a methyl group:

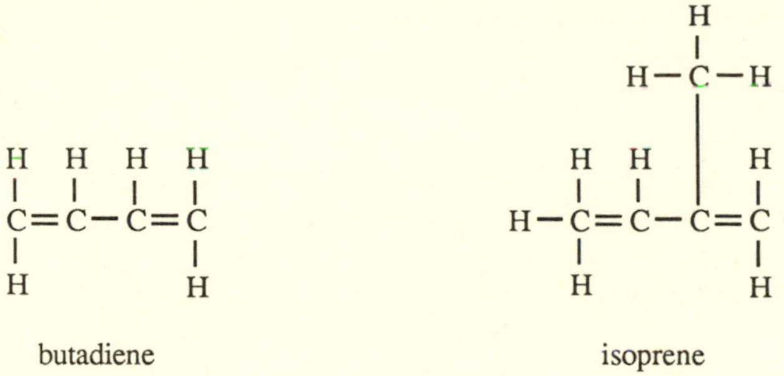

$$
\begin{array}{cccc}
\text{H} & \text{H} & \text{H} & \text{H} \\
| & | & | & | \\
\text{C} & = \text{C} & - \text{C} & = \text{C} \\
| & & & | \\
\text{H} & & & \text{H}
\end{array}
$$

butadiene

$$
\begin{array}{ccc}
& & \text{H} \\
& & | \\
& & \text{H}-\text{C}-\text{H} \\
\text{H} \quad \text{H} & & | \quad \text{H} \\
| \quad | & & | \quad | \\
\text{H}-\text{C}=\text{C} & - & \text{C}=\text{C} \\
| \quad | & & \quad | \\
\text{H} & & \text{H} \quad \text{H}
\end{array}
$$

isoprene

We can polymerize isoprene with sodium, as in the case of buta-diene, but the polyisoprene we obtain is not the same as natural rubber. To explain why, we have to consider three differences between the two polymers. First, isoprene can polymerize with both double bonds, which is called *1,4-addition,* or with only one double bond, which is *1,2* or *3,4* *addition:*

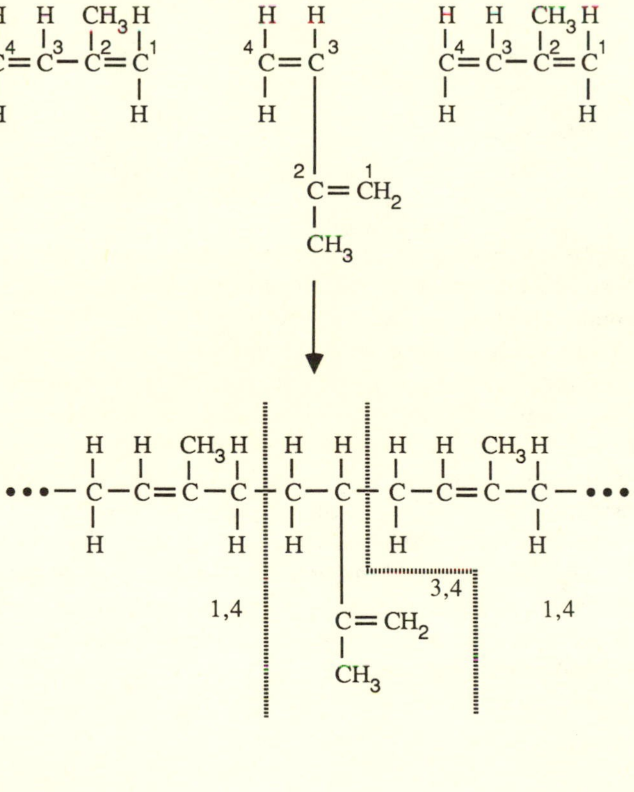

Natural rubber contains only isoprene that had combined in the 1,4 manner, but most synthetic polybutadienes and polyisoprenes contain both 1,2 and 1,4 types of addition. Second, the synthetic polymer can contain chains that are much longer or much shorter than the chains found in natural rubber. Ethyl alcohol always contains two carbon atoms however it is formed, but polyethylene can vary from a few monomer units to several thousands, depending on the conditions used. Third, polyisoprene (or polybutadiene) can exist in two forms. These two forms arise because the double bond is "stiffer" than a single bond. This prevents the two carbon atoms rotating in the axis of the carbon-carbon bond, and if there are two different groups on the adjacent carbon atoms, two distinct forms or *isomers* are created:

$$
\begin{array}{ccc}
\begin{array}{c}
\mathrm{H} \\
| \\
\mathrm{H - C - X} \\
| \\
\mathrm{H - C - X} \\
| \\
\mathrm{H}
\end{array}
& \text{is the same as} &
\begin{array}{c}
\mathrm{H} \\
| \\
\mathrm{H - C - X} \\
| \\
\mathrm{X - C - H} \\
| \\
\mathrm{H}
\end{array}
\end{array}
$$

$$
\begin{array}{ccc}
\begin{array}{c}
\mathrm{H - C - X} \\
\| \\
\mathrm{H - C - X} \\
\text{no rotation}
\end{array}
& \text{is not the same as} &
\begin{array}{c}
\mathrm{H - C - X} \\
\| \\
\mathrm{X - C - H}
\end{array}
\end{array}
$$

$$\text{CIS} \qquad\qquad\qquad \text{TRANS}$$

These two forms are distinguished by the Greek terms *"cis"* (same) and *"trans"* (across). Natural rubber is entirely *cis* polyisoprene and gutta-percha is the *trans* isomer. Synthetic polyisoprene, if prepared by the action of sodium metal, is a mixture of the two types.

It would be natural to assume that polymers are long, stiff rods, but this is not usually the case. If we examine a not-overcrowded string of beads, we notice that the string is fairly flexible. This is also true of most polymer chains. Furthermore, if we drop the string of beads onto a cushion, or stir it in a tub of water, we notice that it forms a coil. Polymer chains also form coils in solution and the length of the molecule is much less than the total length of the polymer chain.

How are these polymer chains formed? If we heat ethyl alcohol with acetic acid (concentrated vinegar) we obtain ethyl acetate, a colorless liquid with an aroma of pears, which is used in nail varnish remover. Similarly, if we heat ethylene glycol (which is used in anti-freeze) with terephthalic acid, we obtain polyethylene terephthalate (*"polyester,"* or Dacron).

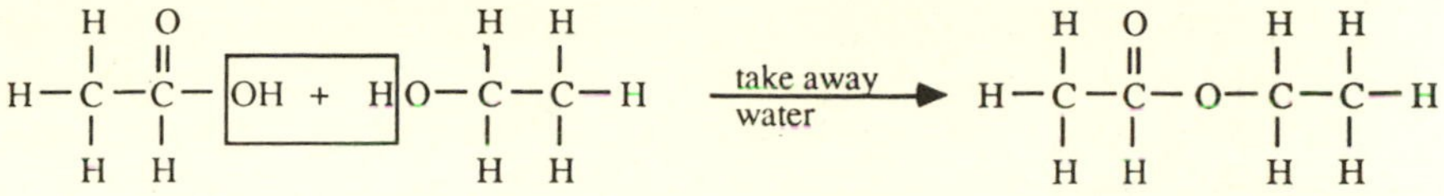

$$
\begin{array}{c}
\underset{\text{acetic acid}}{\text{H}-\overset{\overset{\displaystyle H}{|}}{\underset{\underset{\displaystyle H}{|}}{C}}-\overset{\overset{\displaystyle O}{\|}}{C}-\boxed{\text{OH}}} \;+\; \underset{\text{ethyl alcohol}}{\text{H}\boxed{\text{O}}-\overset{\overset{\displaystyle H}{|}}{\underset{\underset{\displaystyle H}{|}}{C}}-\overset{\overset{\displaystyle H}{|}}{\underset{\underset{\displaystyle H}{|}}{C}}-\text{H}} \;\xrightarrow[\text{water}]{\text{take away}}\; \underset{\text{ethyl acetate}}{\text{H}-\overset{\overset{\displaystyle H}{|}}{\underset{\underset{\displaystyle H}{|}}{C}}-\overset{\overset{\displaystyle O}{\|}}{C}-\text{O}-\overset{\overset{\displaystyle H}{|}}{\underset{\underset{\displaystyle H}{|}}{C}}-\overset{\overset{\displaystyle H}{|}}{\underset{\underset{\displaystyle H}{|}}{C}}-\text{H}}
\end{array}
$$

O|H HO|—C—⟨ ⟩—C—|OH   H|O—C—C—O|H   HO|—C—⟨ ⟩—C—|OH H|O

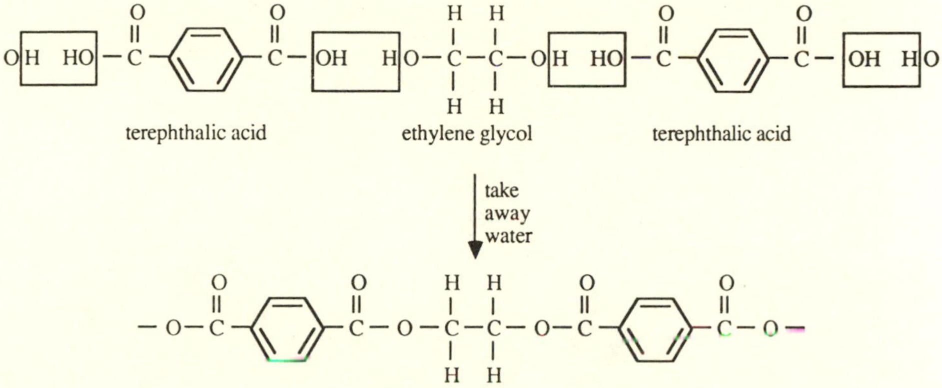

terephthalic acid   ethylene glycol   terephthalic acid

take
away
water

—O—C—⟨ ⟩—C—O—C—C—O—C—⟨ ⟩—C—O—

polyethylene terephthalate

This type of polymer is called a *"condensation polymer."* Nylon is another example of a condensation polymer. In this book, however, we will usually encounter *addition polymers*. Polyethylene and polyisoprene are both examples of addition polymers. This class of polymers is formed in several different ways, but the most common is *free radical polymerization,* which is used to make GR-S. *Free radicals* are not leftists on the loose, but molecules with one atom missing and as a consequence one unshared electron. We usually represent the unshared electron "free" bond by a dot. The free bond in the molecule is very active, more active than the double bond in ethylene.

Free radicals are often formed by the decomposition of unstable molecules. Hydrogen peroxide forms free radicals in the presence of iron compounds:

$$Fe^{++} + H-O-O-H \xrightarrow[\substack{\text{(lawn}\\\text{fertilizer)}}]{\substack{\text{ferrous}\\\text{sulphate}}} H-O^{\cdot} + \overset{\cdot}{O}-H + Fe^{+++}$$

The formation of the hydroxyl radicals from hydrogen peroxide is the *initiation step* in polymerization.

In the second step, this free radical attacks the monomer's double bond

$$HO^\bullet \ + \ CH_2{=}CHX \ \longrightarrow \ HOCH_2{-}\overset{\bullet}{C}HX$$

(X is any suitable atom or group such as:)

$$-\underset{\underset{O}{\|}}{C}-O-CH_3$$

The new free radical rapidly combines with another monomer molecule, in a process called *propagation*.

$$HO-CH_2-CHX-CH_2-\overset{\bullet}{C}HX$$

And so on until the monomer units have formed a long chain:

$$HO-CH_2-CHX-CH_2-CHX-CH_2-CHX-CH_2{=}CHX$$

Of course, *termination* processes intervene sooner or later. Two growing chains are often joined together:

$$-CH_2-CHX-CH_2-CHX-CH_2-CHX-\overset{\bullet}{C}H_2$$

$$+$$

$$\overset{\bullet}{C}HX-CH_2-CHX-$$

$$\downarrow$$

$$-CH_2-CHX-CH_2-CHX-CH_2-CHX-\underset{\underset{CHX-CH_2-CHX-}{|}}{CH_2}$$

Oxygen molecules can often break the chain of free radical additions, and air is usually vigorously excluded from polymerization systems. However, one other type of reaction can occur during free radical polymerization. The free radical abstracts a hydrogen atom from a completed polymer chain:

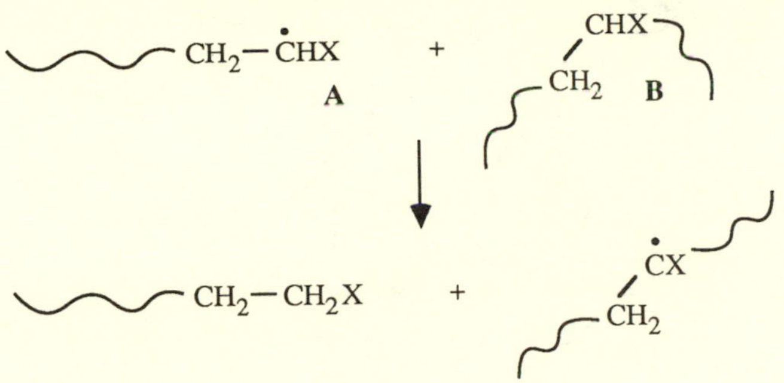

The growing chain (A) is halted, but the active center is transferred to the formerly inert chain (B), which now resumes its growth. Note that the new chain will develop by adding a branch to the old chain:

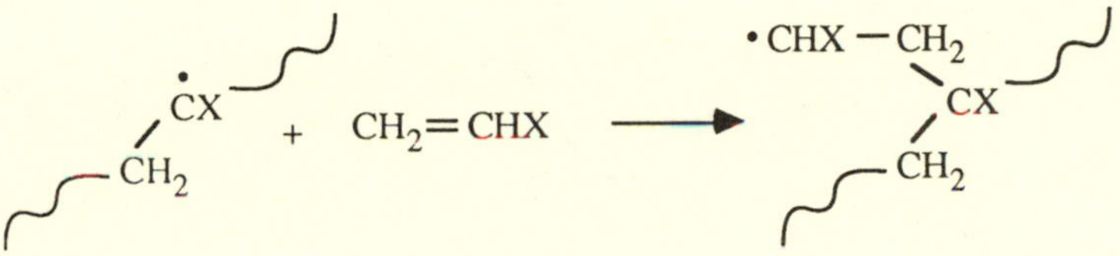

This is called *"branching"* and is common in free radical polymerizations. Polyethylene produced by high pressures contains highly branched chains. The transfer of the active center from one chain to another (*chain transfer*) need not occur directly. A small molecule with a suitable hydrogen atom can effect the transfer:

$$\sim\!\!\!\sim\!\!\!\sim CH_2-\overset{\bullet}{C}HX \; + \; A-H$$

$$\downarrow$$

$$\sim\!\!\!\sim CH_2-CH_2X \; + \; A\bullet$$

(A is a suitable group, for example, $C_{12}H_{25}S-$)

$$A\bullet \; + \; \sim\!\!\!\sim CH_2\!\!\sim\!\!\!\sim^{CHX}\!\!\sim\!\!\!\sim$$

$$\downarrow$$

$$A-H \; + \; \sim\!\!\!\sim CH_2\!\!\sim\!\!\!\sim^{\overset{\bullet}{C}X}\!\!\sim\!\!\!\sim$$

A compound of this type with the correct degree of activity can be used to prevent the formation of excessively long polymer chains. It thereby alters the physical properties of the final polymer and is appropriately called a *"modifier."*

This is essentially how GR-S rubber is formed from butadiene and styrene. Styrene has a similar structure to ethylene but has a phenyl group instead of a hydrogen atom:

```
 H   H                    H   H
 |   |                    |   |
 C = C                    C = C                  styrene
 |   |                    |
 H   H                    H
                        H     C     H
ethylene                 \   / \   /
                          C     C
                          |     ||
                          C     C
                           \   / \
                        H     C     H
                              |
                              H
```

The phenyl group is a very stable symmetrical ring of six carbon atoms and three double bonds. It is also found in benzene (used as a solvent and fuel), phenol (disinfectant), and para-dichlorobenzene (mothkiller, air-freshener).

| styrene | phenol (carbolic acid) | para-dichlorobenzene |

We have already seen it in terephthalic acid. The phenyl group makes styrene easier to polymerize than ethylene. Styrene, in the absence of air, will polymerize if left standing in a bottle, whereas ethylene never polymerizes in ordinary storage. When we polymerize two monomers together, such as butadiene and styrene, we obtain a *copolymer*. In the case of butadiene-styrene copolymer, the distribution of butadiene and styrene units is more or less random along the chain:

(hydrogen atoms removed for clarity)

What is the best way to polymerize butadiene and styrene? A liquid end-product is easier to handle than a solid mass, and the heat produced by the reaction has to be removed efficiently. This rules out the polymerization of butadiene and styrene on their own. An organic solvent could be used, but water is far cheaper. Butadiene and styrene are practically insoluble in water but can be suspended in it in the form of an *emulsion*. Small drops of the two hydrocarbons are held in suspension by an emulsifying agent. Soap and detergents are the most common emulsifiers. (Try making an emulsion by adding a little liquid soap or detergent to a glass of water, stir, then add a little vegetable oil to about one-tenth of the volume of water, and stir vigorously. The milky liquid formed is an emulsion of oil and water.) Emulsion polymerization permits the use of water-soluble initiators such as potassium persulfate, or water-insoluble initiators, like cumene hydroperoxide. Potassium persulfate and cumene hydroperoxide have a weak oxygen-oxygen bond in common with hydrogen peroxide:

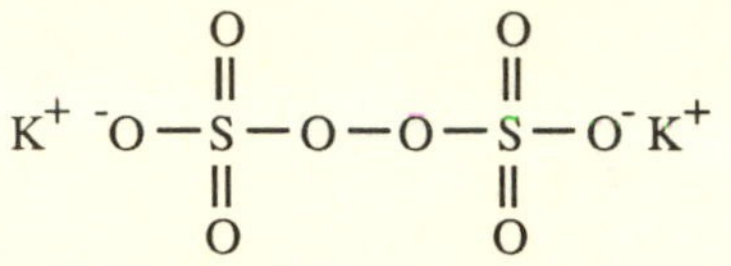

potassium persulphate 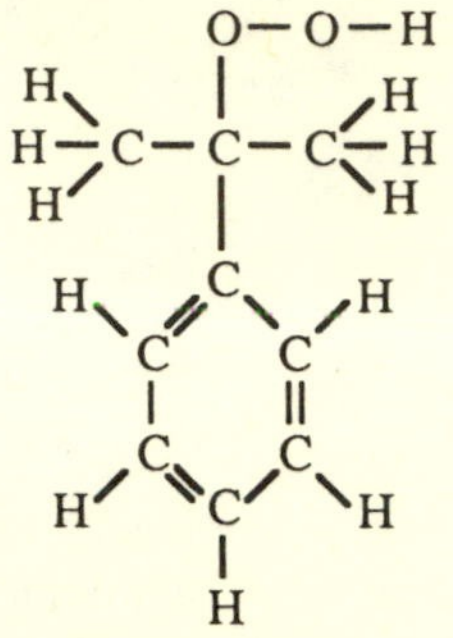   cumene hydroperoxide

In addition to the monomers, the emulsifier, and the initiator, the reaction mixture also contains a modifier to prevent the polymer chains from becoming too long. In the case of butadiene-styrene polymerization, these modifiers belong to a foul-smelling class of compounds called *mercaptans* or thiols. They are similar to alcohols, but contain a sulfur atom instead of an oxygen atom:

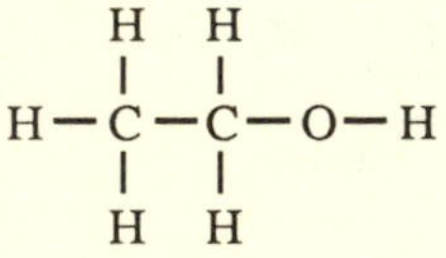

ethyl alcohol

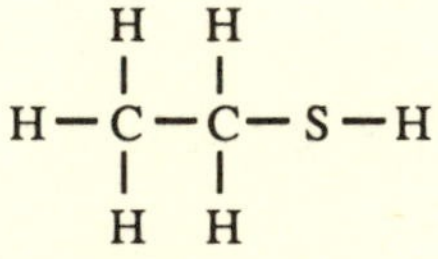

ethyl mercaptan
(ethanethiol)

Added to odorless natural
gas to give it a "stink."

The hydrogen-sulfur bond is weaker than the hydrogen-oxygen bond and is more readily removed by a free radical. Ethyl mercaptan is too active and a mercaptan with more carbon atoms is used, for example, dodecyl mercaptan:

```
      H   H   H   H   H   H   H   H   H   H   H   H
      |   |   |   |   |   |   |   |   |   |   |   |
  H — C — C — C — C — C — C — C — C — C — C — C — C — S — H
      |   |   |   |   |   |   |   |   |   |   |   |
      H   H   H   H   H   H   H   H   H   H   H   H
```

The carbon atoms in dodecyl mercaptan are in a linear sequence; dodecyl mercaptan is an example of a *primary mercaptan*. However, the carbon atoms can be arranged differently; for example, the front carbon atom could link up with three other carbon atoms, and not just one:

```
      H   H   H   H   H   H   H   H   H   CH₃
      |   |   |   |   |   |   |   |   |   |
  H — C — C — C — C — C — C — C — C — C — C — S — H
      |   |   |   |   |   |   |   |   |   |
      H   H   H   H   H   H   H   H   H   CH₃
```

This is still a dodecyl mercaptan, but it is a *tertiary* dodecyl mercaptan. A mixture of tertiary mercaptans is usually employed, and we will just refer to tertiary mercaptans in general.

Before the polymerization gets underway, the initial free radicals react with impurities that have a marked affinity for them, for example, air. This delay is called the *induction period* and can vary from a few seconds to several months, depending on the concentration of these impurities. As the polymer chain grows, it is held in suspension by the emulsifier and forms a very small rubber sphere. The final emulsion of rubber particles is called a latex, as it is similar to the latex ("milk") of the rubber tree. It is then "stripped" of excess butadiene, which is a gas at room temperature, by reducing the pressure to less than a third atmospheric pressure. The latex is then pumped to the top of a tower, where it falls meeting an upward flow of steam that removes the excess styrene. The rubber is then converted into "cream" by the addition of brine. A dilute acid is then added to "cream," which destroys the soap and thereby "cracks" the emulsion, coagulating the rubber as a porous crumb. This crumb is then washed and dried, before being pressed into a block about the size of an average microwave oven (1.6 cubic feet).

There is one complication that we have not yet discussed, the formation of a gel during the polymerization. In contrast to polyethylene, a polymer containing butadiene (or isoprene) still has double bonds. These double bonds can participate in chemical reactions. For example, in *vulcanization* these double bonds react with sulfur to form bridges between the chains:

As we will see, this is a useful application of the double bonds in rubber. Unfortunately, these double bonds can also be attacked by free radicals during the polymerization process:

Cross-links are formed between different chains, and a jelly-like *gel* is produced. The gel soon has a three-dimensional structure that extends throughout the latex particle. If we think of a single polymer chain as a cooked strand of spaghetti, the gel is like a sticky, immobile mass of overcooked spaghetti. In the early stages of polymerization, *gelation* rarely occurs and the formation of polymer chains predominates. As the concentration of monomers falls, however, gelation slowly increases and then suddenly takes over from chain formation at a certain monomer concentration, which is called the "gel point." The gel that is formed during polymerization cannot be larger than the original latex particle and is dispersed within the solid rubber. William O. Baker of Bell Telephone Laboratories called this type of gel *microgel*. A gel can also be formed by heating natural or synthetic rubber in air. The oxygen in air attacks the polymer chains and thereby induces cross-linking:

This gel is not limited in size, like microgel, and can arise in natural rubber as well as GR-S. Baker called this more common gel *macrogel*. While there are important differences between the two types of gel, they are not crucial to our understanding of the chemistry of synthetic rubber, and we will simply refer to gel. More often than not, this will be microgel, but macrogel was also formed during the drying of the rubber crumb, or during the later processing stages.

There are two other methods of addition polymerization, anionic and cationic. The initiator in the case of *anionic polymerization* is an anion, a negatively charged ion. For example, butadiene can be polymerized with butyl lithium:

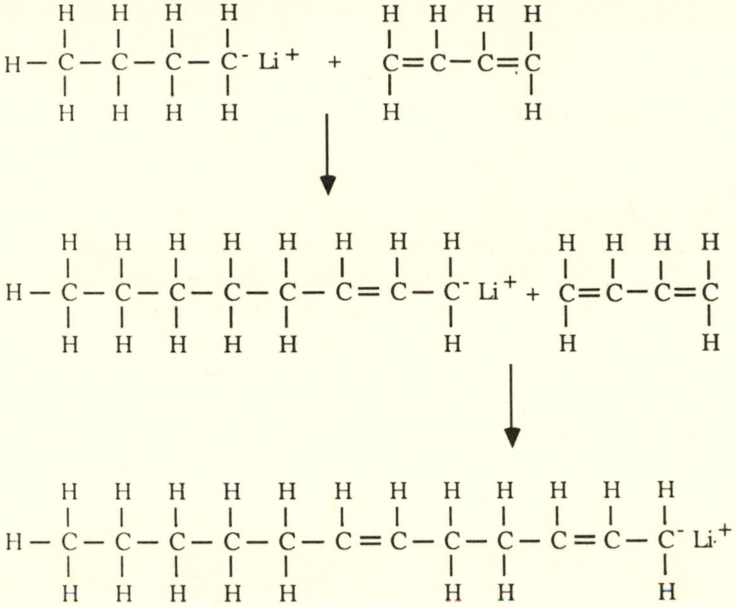

$$
\begin{array}{ccccc}
& H & H & H & H \\
& | & | & | & | \\
H- & C- & C- & C- & C^{-}\ Li^{+} \\
& | & | & | & | \\
& H & H & H & H
\end{array}
\quad + \quad
\begin{array}{cccc}
H & H & H & H \\
| & | & | & | \\
C= & C- & C= & C \\
| & & & | \\
H & & & H
\end{array}
$$

$$
\downarrow
$$

$$
\begin{array}{cccccccc}
H & H & H & H & H & H & H & H \\
| & | & | & | & | & | & | & | \\
H-C= & C- & C- & C- & C- & C= & C- & C^{-}\ Li^{+} \\
| & | & | & | & | & & & | \\
H & H & H & H & H & & & H
\end{array}
\quad + \quad
\begin{array}{cccc}
H & H & H & H \\
| & | & | & | \\
C= & C- & C= & C \\
| & & & | \\
H & & & H
\end{array}
$$

$$
\downarrow
$$

$$
\begin{array}{cccccccccccc}
H & H & H & H & H & H & H & H & H & H & H & H \\
| & | & | & | & | & | & | & | & | & | & | & | \\
H-C= & C- & C- & C- & C- & C= & C= & C- & C- & C= & C- & C^{-}\ Li^{+} \\
| & | & | & | & | & & & | & | & & & | \\
H & H & H & H & H & & & H & H & & & H
\end{array}
$$

By contrast, *cationic polymerization* proceeds with the help of a positively charged ion, or cation. Butyl rubber is produced by the polymerization of isobutylene with boron trifluoride, a strong acid:

$$
BF_3 \ + \ H_2O \ \longrightarrow \ BF_3 OH^- H^+
$$

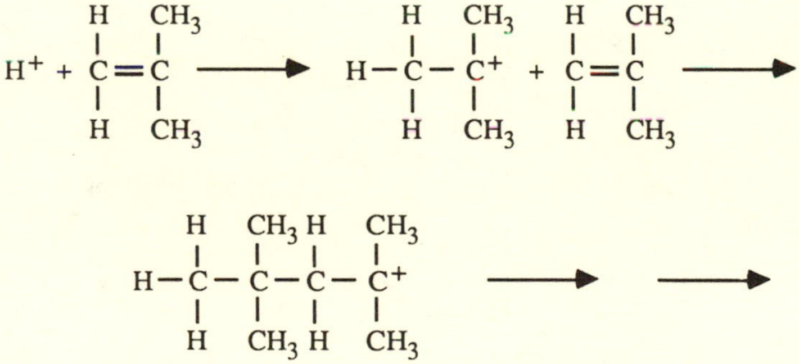

$$
H^+ +
\begin{array}{cc}
H & CH_3 \\
| & | \\
C= & C \\
| & | \\
H & CH_3
\end{array}
\ \longrightarrow \
\begin{array}{cc}
H & CH_3 \\
| & | \\
H-C- & C^+ \\
| & | \\
H & CH_3
\end{array}
\ + \
\begin{array}{cc}
H & CH_3 \\
| & | \\
C= & C \\
| & | \\
H & CH_3
\end{array}
\ \longrightarrow
$$

$$
\begin{array}{cccc}
H & CH_3\ H & CH_3 \\
| & |\quad | & | \\
H-C- & C-C- & C^+ \\
| & |\quad | & | \\
H & CH_3\ H & CH_3
\end{array}
\ \longrightarrow \ \longrightarrow
$$

Boron trifluoride cannot polymerize isobutylene in the absence of water. The actual initiator of the polymerization is a hydrogen ion (proton). Isobutylene has four carbon atoms and one double bond, like butylene, but a different arrangement of the carbon atoms:

isobutylene   butylene

We all know that rubber has different properties from, say, polyethylene or nylon. When we stretch metal wire or a polymer like Silly Putty, it elongates, but when released it retracts only weakly, if at all. By contrast, a rubber band springs back to its original length and shape, more or less. This is a property of *vulcanized rubber*. Dried rubber latex is closer to Silly Putty than a rubber band. You will recall that heating rubber with sulfur produced cross-links between the polymer chains:

When you stretch an ordinary polymer, the polymer coil becomes elongated, and remains stretched when released:

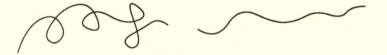

before stretching   during stretching

after stretching

In the case of vulcanized rubber, however, the sulfur cross-links enable the rubber to spring back to its original position:

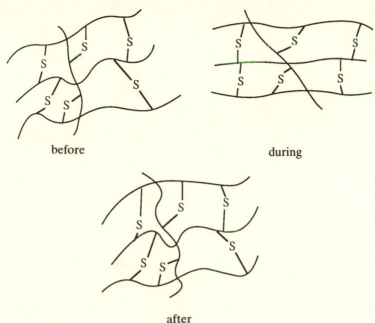

before   during

after

Select Bibliography

Frank Billmeyer. *Synthetic Polymers: Building the Giant Molecule.* Garden City, New York: Doubleday, 1972.

————. *Textbook of Polymer Science.* Third edition. New York: John Wiley, 1984.

Paul J. Flory. *Principles of Polymer Chemistry.* Ithaca, New York: Cornell University Press, 1953.

————. "The Science of Macromolecules." Pages 370–418 in National Research Council, *Outlook for Science and Technology: The Next Five Years.* San Francisco: W. H. Freeman, 1982.

Morris Kaufman. *Giant Molecules: The Technology of Plastics, Fibers and Rubber.* London: Aldus Books, 1968.

Leo Mandelkern. *An Introduction of Macromolecules.* Second edition. New York: Springer-Verlag, 1983.

Herman Mark. *Giant Molecules.* New York: Time-Life Books, 1966.

P. J. T. Morris. *Polymer Pioneers: A Popular History of the Science and Technology of Large Molecules.* Philadelphia: Center for the History of Chemistry, 1986.

M. Morton. *Rubber Technology.* New York: Van Nostrand-Reinhold, 1973.

Kenneth F. O'Driscoll. *The Nature and Chemistry of High Polymers.* New York: Reinhold, 1964.

Raymond B. Seymour and C. E. Carraher, Jr. *Polymer Chemistry: An Introduction.* New York and Basel: Marcel Dekker, 1987.

Works Cited

Interviews

(Affiliations given in parentheses are those at the time of the program.)

Willard C. Asbury and A. Donald Green (Jersey Standard)
> by Peter Morris for the Beckman Center for History of Chemistry (hereafter Beckman Center) oral history program, 9 December 1985.

William J. Bailey (University of Illinois)
> by James J. Bohning for the Beckman Center oral history program, 3 June 1986. Also telephone conversation between W. J. Bailey and Peter Morris, 20 August 1987.

William O. Baker (Bell Telephone Laboratories)
> by Jeffrey L. Sturchio and Marcy Goldstein for Beckman Center and AT&T oral history programs, 23 May 1985 and 18 June 1985.

J. Roger Beatty (Goodrich)
> by Peter Morris, 4 November 1986. Further conversation, 7 August 1987.

Arnold O. Beckman (Beckman Instruments)
> by Jeffrey Sturchio and Arnold Thackray for the Beckman Center oral history program, 23 July 1985.

Harold P. Brown (Goodrich)
> by Colleen Wickey, 12 March 1986.

James D. D'Ianni (Goodyear and ORR)
> by Peter Morris, 5 November 1986. Further conversation, 5 August 1987.

Fred Foster (Firestone)
> by Frank McMillan, 4 September 1973.

Calvin S. Fuller (RRC and Bell Telephone Laboratories)
> by James J. Bohning for the Beckman Center oral history program, 29 April 1986.

Samuel D. Gehman (Goodyear)
> by Peter Morris, 6 November 1986.

Carlin F. Gibbs (Goodrich)
by Peter Morris, 6 August 1987.

Morton Golub (Goodrich)
telephone conversation by Frank McMillan, 29 February 1972.

Paul S. Greer (ORR, OSR)
by Peter Morris for the Beckman Center oral history program, 13 November 1985.

Elbert E. Gruber (Goodrich, General Tire)
by Peter Morris, 5 November 1986.

Samuel E. Horne (Goodrich)
by Frank McMillan, 6 September 1973.

Ben Kastein (Firestone)
by Peter Morris, 21 May 1986.

A. R. Kemp (Bell Telephone Laboratories)
by Herb Elbers for ACS Rubber Division oral history program, 2 April 1965, no transcript.

I. M. Kolthoff (University of Minnesota)
by George Tselos for the Beckman Center oral history program, 15 March 1984.

N. R. Legge (Shell Development)
by Frank McMillan, 8 October 1972.

Herman F. Mark (Brooklyn Polytechnic)
by James Bohning and Jeffrey Sturchio for the Beckman Center oral history program, 3 February 1986, 17 March 1986, and 20 June 1986.

Carl S. Marvel (University of Illinois)
by Leon Gortler and Charles Price for the Beckman Center oral history program, 13 July 1983.

Frank R. Mayo (U.S. Rubber)
by Leon Gortler for the Beckman Center oral history program, 21 January 1981.

John McCool (Goodrich)
by Peter Morris, 6 November 1986.

Maurice Morton (University of Akron)
by Peter Morris, 12 September 1985 and 21 May 1986.

Raymond L. Myers (University of Illinois)
by Peter Morris, 12 December 1985.

Robert Pierson (Goodyear)
by Peter Morris, 6 November 1986.

Charles C. Price (University of Illinois)
by Leon Gortler for the Beckman Center oral history program, 26 April 1979. Further conversation with Peter Morris, 28 May 1987.

E. Ralph Rowzee and Ernest J. Buckler (Polymer)
by Peter Morris, 26 May 1986.

Waldo Semon (Goodrich)
by Herb Elders for ACS Rubber Division oral history program, 16 January 1966.

Richard B. Stambaugh (Goodyear)
  by Peter Morris, 5 November 1986.
Walter Stockmayer (Columbia University)
  by Peter Morris and Jeffrey Sturchio for the Beckman Center oral history
  program, 25 August 1986.
Cheves Walling (University of Chicago and U.S. Rubber)
  by Leon Gortler for the Beckman Center oral history program, 12 September 1979.
Bruno H. Zimm (Columbia University and Brooklyn Polytechnic)
  by James Bohning for the Beckman Center oral history program, 9 September 1986.

Unpublished Documents

(Only those cited more than once are listed here.)
W. O. Baker
  "Discovery and Application of Microgel Molecules in the National Synthetic
  Rubber Program," 13 October 1988. Copy in W. O. Baker Papers, AT&T
  Archives, Warren, New Jersey.
C. S. Fuller
  "History of the Polymer Research Branch, Period December 2, 1942 to
  June 1, 1944," memorandum presented to the 17th meeting of the Rubber
  Research Board, 14 June 1944, National Archives, Washington, D.C., Record
  Group 234, Rubber Reserve Company (hereafter RG 234, RRC)
  Entry 231, PI–173, Minutes of the Rubber Research Board, typescript,
  11 pp.
C. S. Fuller and R. R. Williams
  "Summary of Present Research on Synthetic Rubber," report to the
  first meeting of the Rubber Research Board on 6 January 1943 (but
  clearly written at the end of November 1942), RG 234, RRC, Entry
  231, PI–173, Minutes of the Rubber Research Board, typescript, 18 pp.
Paul S. Greer
  "Cold Rubber," Memorandum RDDR–150 from P. S. Greer to E. D.
  Kelly, 17 February 1954, a copy supplied by Mr. Greer is available at the
  Beckman Center for the History of Chemistry.
Brendan J. O'Callaghan
  "The Government's Rubber Projects (Vol. II)." Prepared 1948; updated
  in 1954 and 1955. RG 234, RRC, Entry 26, PI-173, "Administrative Histories
  of the RFC's Wartime Programs, 1943–54." Photocopy held at the
  Beckman Center for History of Chemistry.
"The University's Synthetic Rubber Program During World War II,"
  November 1956, UI 39/1/8 Box 1, 19.
R. R. Williams
  "Future Research on Synthetic Rubber," 30 June 1943, in casefile 24060–
  1, volume A, AT&T Archives, Warren, New Jersey.

Dissertations

P. Thomas Carroll
"Perspectives on Academic Chemistry in America, 1876–1976: Diversification, Growth and Change," Ph.D. thesis, University of Pennsylvania, 1982. Microfilm number: 83–07294.

Frank R. Chalk
"The United States and the International Struggle for Rubber, 1914–1941," Ph.D. thesis, University of Wisconsin, 1970. Microfilm number: 70–15882.

Jeffrey A. Johnson
"The Chemical Reichsanstalt Association: Big Science in Imperial Germany," Ph.D. thesis, Princeton University, 1980. Microfilm number: 80–08563.

Peter J. T. Morris
"The Development of Acetylene Chemistry and Synthetic Rubber by I.G. Farbenindustrie Aktiengesellschaft: 1926–1945," D.Phil. thesis, Oxford University, 1982. Not available on microfilm, but stored at Beckman Center, and Hagley Library.

Gibson B. Smith
"Rubber for Americans: The Search for an Adequate Supply of Rubber and the Politics of Strategic Materials," Ph.D. thesis, Bryn Mawr College, 1972. Microfilm number: 73–09111.

Jeffrey L. Sturchio
"Chemists and Industry in Modern America: Studies in the Historical Application of Science Indicators," dissertation, University of Pennsylvania, 1981. Microfilm number: 81–27078.

Publications

Basil Achilladelis
"History of UOP: From Petroleum Refining to Petrochemicals." *Chemistry and Industry* (19 April 1975), 377–344.

J. W. Adams and L. H. Howland
"Latex Masterbatching." In Whitby, Davis, and Dunbrook (eds.), *Synthetic Rubber,* 668–681.

Roger Adams
"The Relation of the University Scientists to the Chemical Industries." *Industrial and Engineering Chemistry, News Edition* 13 (1935), 365–367.
"Universities and Industry in Science: Perkin Medal Address." *Industrial and Engineering Chemistry* 46 (1954), 506–510.

Roger Adams, R. C. Fuson, and C. S. Marvel
"The Graduate Training of Chemists." In *Careers in Chemistry and Chemical Engineering.* Washington, D.C.: American Chemical Society, 1951, 31–33.

Hugh Aitken
Syntony and Spark: The Origins of Radio. New York: John Wiley & Sons, 1976.

The Continuous Wave. Princeton, New Jersey: Princeton University Press, 1985.

Hugh Allen
The House of Goodyear: Fifty Years of Men and Industry. Cleveland, Ohio: Corday & Gross, 1949.

David K. Allison
"U.S. Navy Research and Development." In Merritt Roe Smith (ed.), *Military Enterprise and Technological Change.* Cambridge, Massachusetts: MIT Press, 1985, 289–328.

Burton C. Anderson
" 'Speed' Marvel at Du Pont." *Journal of Macromolecular Chemistry—Chemistry* A21 (1984), 1665–1687.

Glenn D. Babcock
History of United States Rubber Company: A Case Study in Corporation Management. Indiana Business Report no. 39, Bureau of Business Research. Bloomington: Indiana University, 1966.

William O. Baker
"Peter Joseph Wilhelm Debye." In W. O. Milligan (ed.), *Proceedings of the Robert A. Welch Conferences on Chemical Research,* volume 20, *American Chemistry—Bicentennial.* Houston: Robert A. Welch Foundation, 1977, 154–199.

F. P. Baldwin
"Modifications of Low Functionality Elastomers." *Rubber Chemistry and Technology* 52 (1979), G77–G84.

John M. Ball
Reclaimed Rubber: The Story of an American Raw Material. New York: Rubber Reclaimers' Association, 1947.

Harry Barron
Modern Rubber Chemistry, second edition. London: Hutchinson's Scientific and Technical Publications, 1947.

James Phinney Baxter III
Scientists Against Time. Boston: Little, Brown, 1946.

Kendall Beaton
Enterprise in Oil: A History of Shell in the United States. New York: Appleton-Century-Crofts, 1959.

R. L. Bebb, E. L. Carr, and L. B. Wakefield
"Synthetic Rubber Polymerization Practices." *Industrial and Engineering Chemistry* 44 (1952), 724–730.

R. L. Bebb and L. B. Wakefield
"German Synthetic-Rubber Developments." In Whitby, Davis, and Dunbrook (eds.), *Synthetic Rubber,* 937–986.

F. A. Bovey, I. M. Kolthoff, A. I. Medalia, and E. J. Meehan
Emulsion Polymerization, High Polymers volume IX. New York: Interscience, 1955.

Stanley E. Boyle
"Government Promotion of Monopoly Power: An Examination of the Sale of the Synthetic Rubber Industry." *Journal of Industrial Economics* 9 (1961), 151–169.

Robert Brasted
Interview of I. M. Kolthoff for "Impact." *Journal of Chemical Education* 50 (1973), 663–666.

E. J. Buckler
"Canadian Contributions to Synthetic Rubber Technology." *Canadian Journal of Chemical Engineering* 62 (1984), 3–12.

Bunawerke Hüls GmbH
Buna: Dokumente uber ein neues Werk zum beginn der Produktion am 15. September 1958. Bunawerke Hüls, Marl, West Germany, 1958.

C. E. Carraher, Jr.
"Polymer Education and the Mark Connection." In Stahl (ed.), *Polymer Science Overview,* 123–142.

Rexmond C. Cochrane
Measures for Progress: A History of the National Bureau of Standards. NBS Miscellaneous Publication 275, National Bureau of Standards, Department of Commerce, Washington, D.C., 1966. Rprt. New York: Arno Press, 1976.

James B. Conant, Karl T. Compton, and Bernard M. Baruch
Report of the Rubber Survey Committee. Washington, D.C.: Government Printing Office, 1942.

M. M. Crow
"Synthetic Fuel Technology Nondevelopment and the Hiatus Effect: The Implications of Inconsistent Public Policy." In E. J. Yanarella and W. C. Green (eds.), *The Unfulfilled Promise of Synthetic Fuels: Technological Failure, Policy Immobilism or Commercial Illusion,* Contributions in Political Science no. 179. Westport, Connecticut: Greenwood Press, 1987, 33–56.

Eli M. Dannenberg
"The Carbon Black Industry: Over a Century of Progress." In *Rubber Division 75th Anniversary,* Rubber World Magazine, Akron, 1984, 35–40.

Carroll C. Davis and John T. Blake (eds.)
The Chemistry and Technology of Rubber. American Chemical Society Monograph Series. New York: Reinhold for the ACS Rubber Division, 1937.

James D. D'Ianni
"Fun and Frustrations with Synthetic Rubber." *Rubber Chemistry and Technology* 50 (1977), G67–G77.

David Dietz
The Goodyear Research Laboratory. Akron, Ohio: Goodyear, 1943.

R. P. Dinsmore and R. D. Juve
"The Processing and Compounding of GR-S." In Whitby, Davis, and Dunbrook (eds.), *Synthetic Rubber,* 373–424.

R. F. Dunbrook
"Historical Review." In Whitby, Davis, and Dunbrook (eds.), *Synthetic Rubber,* 32–55.

H. F. Eicke
"Peter J. W. Debye's Beiträge zur Makromolekularen Wissenschaft—ein Beispiel zukunftsweisender Forschung." *Chimia* 38 (1984), 347–353.

F. R. Eirich (ed.)
Science and Technology of Rubber. New York: Academic Press, 1978.
John L. Enos
Petroleum, Progress and Profits: A History of Process Innovation. Cambridge, Massachusetts: MIT Press, 1962.
M. Fagen (ed.)
A History of Engineering and Science in the Bell System: The Early Years, 1875–1925. Murray Hill, New Jersey: Bell Telephone Laboratories, 1975.
Donald D. Fitts
"John Gamble Kirkwood." In Miles (ed.), *American Chemists and Chemical Engineers.*
Frederick M. Fowkes
"William Draper Harkins." In *Dictionary of Scientific Biography,* volume 6. New York: Charles Scribner's Sons, 1972, 117–119.
Christopher Freeman
An Introduction to the Economics of Industrial Innovation, second edition. Cambridge, Massachusetts: MIT Press, 1982.
M. J. French
"The Emergence of a U.S. Multinational Enterprise: The Goodyear Tire and Rubber Company, 1910–1939." *Economic History Review,* second series, 40 (1987), 64–79.
Charles F. Fryling
"Emulsion Polymerization Systems." In Whitby, Davis, and Dunbrook (eds.), *Synthetic Rubber,* 224–288.
C. S. Fuller
"Some Recent Contributions to Synthetic Rubber Research." *Bell System Technical Journal* 25 (1946), 351–384.
William C. Geer
The Reign of Rubber. New York: Century, 1922.
August W. Giebelhaus
Business and Government in the Oil Industry: A Case Study of Sun Oil, 1876–1945. Industrial Development and the Social Fabric no. 5. Greenwich, Connecticut: JAI Press, 1980.
Sheldon Glashow and Leon Lederman
"The SSC: A Machine for the Nineties." *Physics Today* 38 (March 1985), 29–37.
"Government Laboratories of the Office of Rubber Reserve." *Chemical and Engineering News* 24 (1946), 3024–3026.
E. Guth
"Birth and Rise of Polymer Science—Myth and Truth." *Journal of Applied Polymer Science: Applied Polymer Symposium* 35 (1979), 1–12.
P. W. Hamlett
"Technological Policy Making in Congress: The Creation of the U.S. Synthetic Fuels Corporation." In E. J. Yanarella and W. C. Green (eds.), *The Unfulfilled Promise of Synthetic Fuels: Technological Failure, Policy Immobilism or Commercial Illusion.* Contributions in Political Science no. 179. Westport, Connecticut: Greenwood Press, 1987, 53–69.

Williams Haynes and Ernst A. Hauser
Rationed Rubber and What To Do About It. New York: Knopf, 1942.

Donald R. Hays
"Neil Elbridge Gordon." In Miles (ed.), *American Chemists and Chemical Engineers.*

Vernon Herbert and Attilio Bisio
Synthetic Rubber: A Project That Had to Succeed, Contributions in Economics and Economic History no. 63. Westport, Connecticut: Greenwood Press, 1985.

Claus Heuck
"Ein Beitrag zur Geschichte der Kautschuk-Synthese: Buna-Kautschuk I.G. (1926–1945)." *Chemiker-Zeitung* 94 (1970), 147–157.

Richard G. Hewlett and Oscar E. Anderson, Jr.
A History of the United States Atomic Energy Commission, vol. 1, *The New World, 1939–1946.* College Park, Pennsylvania: Pennsylvania State University Press, 1962.

Friedrich Hölscher
Kautschuk, Kunststoffe, Fasern. Schriftenreihe des Firmenarchivs des BASF AG no. 10. Ludwigshafen am Rhein, West Germany: BASF, 1972.

Samuel E. Horne
"Polymerization of Diene Monomers by Ziegler Type Catalysts." *Rubber Chemistry and Technology* 53 (1980), G68–G79.
"The History of Synthetic Rubber." In *Rubber Division 75th Anniversary,* Rubber World Magazine, Akron, 1984, 5–7.

David A. Hounshell and John K. Smith
Science and Corporate Strategy: Du Pont R&D, 1902–1980. New York: Cambridge University Press, 1988.

Frank A. Howard
Buna Rubber: The Birth of an Industry. New York: Van Nostrand, 1947.

University of Illinois
Chemistry: 1941–1951. Urbana: University of Illinois, 1951.

John Jewkes, David Sawers, and Richard Stillerman
The Sources of Invention, second edition. London: Macmillan, 1969.

Warren C. Johnson
"William Draper Harkins." In Miles (ed.), *American Chemists and Chemical Engineers.*

Daniel P. Jones
"Julius Oscar Stieglitz." In Miles (ed.), *American Chemists and Chemical Engineers.*

George B. Kauffmann
"William Draper Harkins (1873–1951): A Controversial and Neglected American Physical Chemist." *Journal of Chemical Education* 62 (1985), 758–761.

Patrick Kelly and Melvin Kranzberg (eds.)
Technological Innovation: A Critical Review of Current Knowledge. San Francisco: San Francisco Press, 1978.

Milton Kerker
"Classics and Classicists of Colloid and Interface Science. 2. John William Strutt, Lord Rayleigh." *Journal of Colloid and Interface Science* 113 (1986), 589–593.

Erich Konrad
"Uber die Entwicklung des synthetischen Kautschuks in Deutschland." *Angewandte Chemie* 62 (1950), 491–496.

E. Konrad and W. Becker
"Zur Geschichte des bei tiefer Temperatur polymerisierten synthetischen Kautschuks." *Angewandte Chemie* 62 (1950), 423–426.

Clayton R. Koppes
JPL and the American Space Program: A History of the Jet Propulsion Laboratory. New Haven: Yale University Press, 1982.

Melvin Kranzberg
"Science-Technology and Warfare: Action, Reaction and Interaction in the Post-World War II Era." In M. D. Wright and L. J. Paszek (eds.), *Science, Technology and Warfare.* Washington, D.C.: Government Printing Office for Office of Air Force History and U.S. Air Force Academy, 1971, 146–151.

Paul Kränzlein
Chemie im Revier—hüls. Dusseldorf and Vienna: Econ Verlag, 1980.

H. A. Laitinen and E. J. Meehan
"Happy Birthday I. M. Kolthoff." *Analytical Chemistry* 56 (1984), 248A–262A.

J. Langrish, M. Gibbons, W. G. Evans, and F. R. Jevons
Wealth from Knowledge: A Study of Innovation in Industry. London: Macmillan, 1972.

Henrietta Larson, Evelyn Knowlton, and Charles Popple
New Horizons, 1927–1950. New York: Harper & Bros., 1971.

Henrietta Larson and Kenneth Porter
History of Humble Oil & Refining Company: A Study in Industrial Growth. New York: Harper & Bros., 1959.

Stuart W. Leslie
Boss Kettering. New York: Columbia University Press, 1983.

Alfred Lief
The Firestone Story. New York: Whittlesey House, 1951.

J. J. Lingane
"Izaak Maurits Kolthoff." *Talanta* 11 (1964), 67–73.

John W. Livingston and John T. Cox, Jr.
"The Manufacture of GR-S." In Whitby, Davis, and Dunbrook (eds.), *Synthetic Rubber,* 175–223.

Heino Logemann and Gottfried Pampus
"Buna S—Seine grosstechnische Herstellung und seine Weiterentwicklung—ein geschichtlicher Überblick." *Kautschuk und Gummi, Kunststoffe* 23 (1970), 479–486.

John Logsdon
The Decision to Go to the Moon, Project Apollo and the National Interest. Cambridge, Massachusetts: MIT Press, 1970.

Milton Lomask
   A Minor Miracle: An Informal History of the National Science Foundation.
   Washington, D.C.: Government Printing Office for National Science Foundation, 1976.
Leo Mandelkern, James E. Mark, Ulrich W. Suter, and Do Y. Yoon
   Selected Works of Paul J. Flory, three volumes. Stanford, California: Stanford University Press, 1985.
Edwin Mansfield
   The Economics of Technological Change. New York: Norton, 1968.
Hans Mark and Arnold Levine
   The Management of Research Institutions: A Look at Government Laboratories. Washington, D.C.: Scientific and Technical Information Branch, National Aeronautics and Space Administration, 1984.
Herman F. Mark
   "Polymers—Past, Present, Future." In W. O. Milligan (ed.), *Proceedings of The Robert A. Welch Foundation Conferences on Chemical Research,* volume 10, *Polymers.* Houston: Robert A. Welch Foundation, 1967, 19–55.
   "Polymer Chemistry: The Past Hundred Years." *Chemical and Engineering News* 54 (6 April 1976), ACS Centennial Issue, 176–189.
   "Polymer Chemistry in Europe and America—How It All Began." *Journal of Chemical Education* 58 (1981), 527–534.
   "The Contribution of Carl (Speed) Marvel to Polymer Science." *Journal of Macromolecular Science—Chemistry* A21 (1984), 1567–1606.
C. S. Marvel
   "The Development of Polymer Chemistry in America: The Early Days." *Journal of Chemical Education* 58 (1981), 535–539.
C. S. Marvel and Henry E. Baumgarten
   "Chemical Study of the Structure of Diene Polymers and Copolymers." In Whitby, Davis, and Dunbrook (eds.), *Synthetic Rubber,* 289–315.
Frank R. Mayo
   "The Discovery of the Peroxide Effect." In Waters (ed.), *Vistas in Free Radical Chemistry,* 139–142.
   "The Evolution of Free Radical Chemistry at Chicago." *Journal of Chemical Education* 63 (1987), 97–99.
Frank R. Mayo and Cheves Walling
   "Copolymerization." *Chemical Reviews* 46 (1950), 190–287.
D. McIntyre and F. Gornick (eds.)
   Light Scattering from Dilute Polymer Solutions. International Science Review volume 3. New York: Gordon & Breach, 1964.
Frank M. McMillan
   The Chain Straighteners. Fruitful Innovation: The Discovery of Linear and Stereoregular Synthetic Polymers. London: Macmillan, 1979.
Thomas Midgley, Jr.
   "Synthetic and Substitute Rubbers." In Davis and Blake (eds.), *The Chemistry and Technology of Rubber,* 677–704.

Wyndham D. Miles
"Wilder Dwight Bancroft." In Miles (ed.), *American Chemists and Chemical Engineers.*

Wyndham D. Miles (ed.)
American Chemists and Chemical Engineers. Washington, D.C.: American Chemical Society 1976.

S. Millman (ed.)
A History of Engineering and Science in the Bell System. Physical Sciences (1925–1980). Murray Hill, New Jersey: AT&T Bell Telephone Laboratories, 1983.

Herbert Morawetz
Polymers: The Origins and Growth of a Science. New York: Wiley-Interscience, 1985.

Peter J. T. Morris
"Buna S Versus GR-S: A Comparative Study of Industrial Research in Germany and the United States." A paper presented at the HSS-BSHS joint meeting in Manchester, U.K., July 1988.

Maurice Morton
"History of Synthetic Rubber." In Raymond B. Seymour (ed.), *History of Polymer Science and Technology,* 225–239.
"Rubber Enters the Polymer Age." *Rubber Chemistry and Technology* 58 (1985), G75–G90.

Maurice Morton (ed.)
Introduction to Rubber Technology. New York: Reinhold, 1959.

Donald M. Nelson
Arsenal of Democracy: The Story of American War Production. New York: Harcourt Brace, 1946.

Richard R. Nelson and Richard N. Langlois
"Industrial Innovation Policy: Lessons from American History." *Science* 219 (1983), 814–818.

Richard R. Nelson and Sidney G. Winter
An Evolutionary Theory of Economic Change. Cambridge, Massachusetts: Belknap Press of Harvard University, 1982.

Maurice O'Reilly
The Goodyear Story. Elmsford, New York: Benjamin, 1983.

Charles F. Phillips, Jr.
Competition in the Synthetic Rubber Industry. Chapel Hill, North Carolina: University of North Carolina Press, 1963.

Phillips Petroleum
Phillips: The First 66 Years. Bartlesville, Oklahoma: Phillips Petroleum, 1983.

Charles Popple
Standard Oil Company (New Jersey) in World War II. New York: Standard Oil, 1952.

E. J. Powers and G. A. Serad
"History and Development of Polybenzimidazoles." In Raymond B. Seymour and Gerald S. Kirshenbaum (eds.), *High Performance Poly-*

mers: Their Origin and Development. New York: Elsevier, 1986, 355–373.

Charles C. Price
"How Chemists Create a New Product." *The Chemist* 38 (1961), 131–132.

Proceedings, Joint Army-Navy-Air Force Conference on Elastomer Research and Development. Washington, D.C.: National Academy of Sciences–National Research Council publication no. 30, 1954.

Yakov M. Rabkin
"Technological Innovation in Science: The Adoption of Infrared Spectroscopy by Chemists." *Isis* 78 (1987), 31–54.

Leonard S. Reich
The Making of American Industrial Research: Science and Business at GE and Bell, 1876–1926. Cambridge, U.K.: Cambridge University Press, 1985.

Nathan Rosenberg
Perspectives On Technology. Cambridge, U.K.: Cambridge University Press, 1976.
Inside the Black Box: Technology and Economics. Cambridge, U.K.: Cambridge University Press, 1982.

Davis R. B. Ross
"Patents and Bureaucrats: U.S. Synthetic Rubber Developments Before Pearl Harbor." In Joseph R. Frese, S. J., and Jacob Judd (eds.), *Business and Government.* Tarrytown, New York: Sleepy Hollow Press and Rockefeller Archive Center, 1985, 119–155.

"Rubber—How Do We Stand?" *Fortune* 25 (June 1942), 94–96, 192–194.

Harvey M. Sapolsky
"Academic Science and the Military: The Years Since the Second World War." In Nathan Reingold (ed.), *The Sciences in the American Context: New Perspectives.* Washington, D.C.: Smithsonian Institution Press, 1979, 379–399.

Jacob Schmookler
Invention and Economic Growth. Cambridge, Massachusetts: Harvard University Press, 1966.

Herman E. Schroeder
"Facets of Innovation." *Rubber Chemistry and Technology* 57 (1984), G86–G106.

Waldo L. Semon
"Thirty Years' Contributions to the Science of Synthetic Rubber." *Chemical and Engineering News* 21 (1943), 1613–1619.

John W. Servos
"A Disciplinary Program That Failed: Wilder D. Bancroft and the *Journal of Physical Chemistry* 1896–1933." *Isis* 73 (1982), 207–232.

Raymond B. Seymour (ed.)
History of Polymer Science and Technology. New York and Basel: Marcel Dekker, 1982.

John K. Smith
"The Ten Year Invention: Neoprene and Du Pont Research, 1930–1939." *Technology and Culture* 26 (1985), 34–55.

Charles P. Smyth
"Debye, Peter Joseph William." In *Dictionary of Scientific Biography,* volume 3. New York: Charles Scribner's Sons, 1971, 617–621.

Robert Solo
"The Sale of the Synthetic Rubber Plants." *The Journal of Industrial Economics* 2 (1953), 32–43.
"Research and Development in the Synthetic Rubber Industry." *Quarterly Journal of Economics* 68 (1954), 61–82.
Synthetic Rubber: A Case Study in Technological Development Under Government Direction. Study No. 18 for the Sub-Committee on Patents, Trademarks and Copyright, Committee on the Judiciary, U.S. Senate, 85th Cong. 2nd sess., 1959, Committee Print, 93. Reprinted as *Across the High Technology Threshold: The Case of Synthetic Rubber.* Norwood, Pennsylvania: Norwood Editions, 1980.

Peter H. Spitz
Petrochemicals, The Rise of an Industry. New York: Wiley-Interscience, 1988.

B. Sheldon Sprague
"An Industrial Innovation That Was Nearly Shelved." *Research Management* (May–June 1986), 26–29.

G. Allan Stahl
"Herman F. Mark: The Geheimrat." In Stahl (ed.), *Polymer Science Overview,* 61–88.
"Development of Modern Polymer Theory." *CHEMTECH* 14 (1984), 492–495.

G. Allan Stahl (ed.)
Polymer Science Overview: A Tribute to Herman F. Mark. ACS Symposium Series no. 175. Washington, D.C.: American Chemical Society, 1981.

Walter H. Stockmayer
"The 1974 Nobel Prize for Chemistry." *Science* 186 (1974), 724–726.

Walter H. Stockmayer and Bruno H. Zimm
"When Polymer Science Looked Easy." *Annual Review of Physical Chemistry* 35 (1984), 1–21.

Jeffrey L. Sturchio
"Chemistry and Corporate Strategy at Du Pont." *Research Management* 27 (1984), 10–18.

Nuala Swords-Isherwood
The Process of Innovation. London: British-North American Committee, 1984.

D. Stanley Tarbell and Ann Tracy Tarbell
Roger Adams: Scientist and Statesman. Washington, D.C.: American Chemical Society, 1981.

D. S. Tarbell, Ann T. Tarbell, and R. M. Joyce
"The Students of Ira Remsen and Roger Adams." *Isis* 71 (1980), 620–626.

Arnold Thackray
"University-Industry Collaboration and Chemical Research: A Historical Perspective." In *University-Industry Research Relationships,* Report of the

National Science Board of the National Science Foundation. Washington, D.C.: Government Printing Office, 1982, 193–223.

Robert M. Thomas
"Early History of Butyl Rubber." *Rubber Chemistry and Technology* 42 (1969), G90–G96.

Erik G. M. Tornqvist
"Polyolefin Elastomers—Fifty Years of Progress." In R. B. Seymour and Tai Cheng (eds.), *History of Polyolefins: The World's Most Widely Used Polymers*. Dordrecht, Netherlands: D. Reidel, 1986, 143–161.

William M. Tuttle, Jr.
"The Birth of an Industry: The Synthetic Rubber 'Mess' in World War II." *Technology and Culture* 22 (1981), 35–67.

R. D. Ulrich (ed.)
Contemporary Topics in Polymer Science, volume 1, *Macromolecular Science: Retrospect and Prospect*. New York and London: Plenum Press, 1978.

U.S. Congress, Joint Economic Committee
Research and Innovation: Developing a Dynamic Nation. Special Study on Economic Change, volume three, 96th Cong., 2nd sess., 1980, Committee Print.

H. S. van Klooster
"Hugo Rudolph Kruyt." *Journal of Chemical Education* 19 (1942), 165.

Cheves Walling
"The Contributions of Morris S. Kharasch to Polymer Science." In Waters (ed.), *Vistas in Free Radical Chemistry*, 143–150.
"Forty Years of Free Radicals." In W. A. Pryor (ed.), *Organic Free Radicals*. ACS Symposium Series no. 69. Washington, D.C.: American Chemical Society, 1978, 1–11.
"The Development of Free Radical Chemistry." *Journal of Chemical Education* 63 (1986), 99–102.

W. C. Warner
"Arthur Edgar Juve." In Miles (ed.), *American Chemists and Chemical Engineers*.

W. A. Waters (ed.)
Vistas in Free Radical Chemistry: In Memoriam, Dr. Morris S. Kharasch. London: Pergamon Press, 1959.

W. A. Waters and F. R. Mayo
"The Significance of the Work of M. S. Kharasch in the Development of Free-Radical Chemistry." In Waters (ed.), *Vistas in Free Radical Chemistry*, 1–5.

Frank H. Westheimer
"Morris Selig Kharasch." *Biographical Memoirs of the National Academy of Sciences* 34 (1960), 123–134.

G. S. Whitby, C. C. Davis, and R. F. Dunbrook (eds.)
Synthetic Rubber. New York: John Wiley & Sons; and London: Chapman & Hall, 1954.

J. L. White
"Rheological Behavior of Unvulcanized Rubber." In Eirich (ed.), *Science and Technology of Rubber*, 223–289.

Don Whitehead
The Dow Story: The History of the Dow Chemical Company. New York: McGraw-Hill, 1968.

Charles Morrow Wilson
Trees and Test Tubes: The Story of Rubber, New York: H. Holt, 1943.

George Wise
Willis R. Whitney and the Rise of American Industrial Research. New York: Columbia University Press, 1985.

Howard Wolf
The Story of Scrap Rubber. Akron, Ohio: A. Schulman, 1943.

Howard Wolf and Ralph Wolf
Rubber: A Story of Glory and Greed. New York: Covici Friede, 1936.

Lawrence A. Wood
"Physical Chemistry of Synthetic Rubbers." In Whitby, Davis, and Dunbrook (eds.), *Synthetic Rubber,* 316–372.

"Memories of Synthetic Rubber at the National Bureau of Standards: 1936–1956." In *World War II Synthetic Rubber Program: Mission, Record, Mechanisms, Significance and Its Messages for Today.* Typescript published by the Washington Rubber Group and Rubber Division, American Chemical Society, Washington, D.C., 1979.

Franz I. Wünsch
"Das Werk Hüls. Geschichte der Chemische Werke Hüls AG in Marl von 1938 bis 1949." *Tradition* 9 (1964), 70–79.

Subject Index

Activators: diazothioethers, 38, 55, 81; early formulae, 33; ferricyanides, 33–34, 38, 81; Iron (II), 35, 38, 79, 81–83, 132

Analysis, 141; of reactants, 15, 79, 80–81, 83; of rubber, 104, 110, 112; of styrene content, 15–16, 81, 104, 105, 111

Atomic bomb. *See* Manhattan Project *in name index*

Cable and wire insulation, 84, 102, 103, 104, 114

Carbon black, 17, 32, 36, 38, 51

Chain transfer. *See* Polymerization, free radical

Chemical structure of GR-S: by infra-red spectrophotometry, 40–41, 45; by oxidation, 73; by ozonolysis, 72–73; by ultra-violet spectrophotometry, 81

Chemistry and Technology of Rubber, 87

Cold War, 16, 54

Configuration of polymers, 85–86, 103, 105, 132

Copolymerization. *See* Polymerization, copolymerization

Depolymerization, 112

Disposal of plants, 19, 21–22, 47, 53–54

Education, 71–72, 133–34

Elasticity, 131

Electrochemical analysis, 30, 79–80

Emulsions. *See* Polymerization, emulsion

Evaluation of new polymers, 21, 110, 113

Extenders, 18–19, 36–38. *See also* Rubber, oil-extended, *in polymer index*

Fatty acids. *See* Soap

Free radicals. *See* Polymerization, free radical

Gelation, 14, 103, 130, 157–58. *See also* Microgel

Industry-university collaboration, 52, 88, 141; achievements of, 2–3, 20–21, 55; attitude of the rubber companies, 90; Chicago and U.S. Rubber, 77, 78, 91; Debye, 84–85; Illinois, 31, 70–71; Kolthoff and the rubber companies, 31, 82, 90–91; pre-war, 69–70; Price and General Tire, 51

Initiators: peroxide, 33, 34–35, 51, 79, 81–83, 91, 114, 132, 149–50, 154; persulfate, 29, 80, 154

Innovation, 27, 50–59, 88–92; incremental, 58–59, 142; radical, 58–59, 88–89, 142. *See also* Research

Journal of Chemical Education, 88

Kinetic studies, 15, 31, 77–78, 80–81, 83, 127–29, 131–32, 141
Korean War, 17, 19

Light scattering, 16, 84–85, 98 n.89, 127, 136, 141

Mass spectrometry, 112
Masterbatching of GR-S, 38
Mechanistic studies. *See various entries under* Polymerization
Mercaptans. *See* Modifiers
Microgel, 105–9, 130, 136, 157
Modification (including modifier disappearance), 15, 30–31, 77, 80–81, 104
Modifiers, 14, 36, 37, 39, 77–78, 90, 152, 154; di-isopropylxanthogen disulfide (Dixie), 30, 33; primary (Lorol, DDM), 15, 27, 29–31, 33, 35, 72, 80–81, 155; synthesis of new, 30, 72, 104; tertiary (sulfole), 15, 17, 30–31, 35, 51, 80–81, 155
Molecular weights, 85, 108; definition of, 98 n.101. *See also* Light scattering; Viscosity
Monomers, synthesis of new, 8, 18, 21, 73–75, 113

Nuclear power. *See* Manhattan Project *in name index*
Nuclear submarine program, 53, 59

Petroleum industry, 1, 5 n.1, 14, 40, 67 n.131. *See also under individual companies in name index*
Physical chemistry of polymers, 83–86, 103, 105–9, 112, 130–32
Physical structure of GR-S, 34, 131; *cis/ trans* studies, 41, 42, 44, 47, 48
Physical testing of rubbers, 28, 42, 47, 106–7, 110, 111–12, 113, 130–31
Polarography (by Kolthoff and Lingane), 80
Polymerization: Alfin, 20–21, 43, 47, 90–91; alkali metal, 20–21, 42–43, 43–44, 47, 73, 113–14, 146–48; anionic, 43–44, 129, 158–59; cationic, 49, 129, 159; condensation, 48, 103, 130, 148–49; copolymerization, 127–29, 153;

emulsion, 16, 77, 78, 80–81, 129, 153–55; free radical, 30, 76–79, 80–81, 94 n.44, 112, 131–32, 149–58; inhibition of, 14–15, 31–32, 72, 76, 150; photo-, 78; redox (low temperature), 17, 32–35, 38–39, 57, 74–77, 79, 81–83, 114, 131–32; stereoregular, 20, 42–49; vinyl, 71; Ziegler-Natta, 20, 45–46, 49, 88–89. *See also* Depolymerization; Kinetic studies
Principles of Polymer Chemistry (Flory), 86, 124

Quality control, 101, 105–6, 107, 111–12, 113, 117–18

Recipes: Custom, 17, 35, 55, 82, 114; Goodrich (H Formula), 33, 61 n.32; Mutual, 29, 33; peroxamine, 55, 82, 114; sulfoxylate, 82; Veroxasulfide, 55, 82
Research: generic, 115–18; military-funded, 18, 48, 54, 89 , 114, 132–33, 136, 141; mission-oriented, 51–52, 89. *See also* Innovation
Rheology, 111–12, 130–31
Rubber Act (1948), 29
Rubber cartels, 1, 46
Rubber Producing Facilities Disposal Act (1953), 19, 21
Rubber research program: achievements of, 2–3, 20–21, 55, 136, 141; criticisms of, 3–4, 50–55, 136–37, 141; direction of, 12–13, 16, 21–22, 51–54, 70; funding of, 53, 54

Soap, 14, 15, 29, 30, 31, 34, 81; rosin, 31–32, 35, 51, 81, 82, 91
Solar cells, 66 n.122
Space program, 57, 67 n.131, 127, 142. *See also* National Aeronautics and Space Administration *in name index*
Spectrophotometry, infra-red, 39–41, 44, 45, 47, 63 n.77, 63 n.80, 127; ultraviolet, 81
Styrene, replacement of. *See* Monomers, synthesis of
Styrene content, measurement of, 15–16, 74, 81, 104, 105, 111

Synthetic fuels, 2, 59, 66 n.123
Synthetic Rubber, 87, 115

Textbook of Polymer Chemistry (Bill-
  meyer), 124
Thermal degradation (plasticization) of
  GR-S, 36, 107–8, 113–14
Tire tests, 110, 113

Udex press for styrene, 75

Viscosity: measurement of, 28, 106, 108,
  111–12, 117, 130; regulation of, 15,
  30
Vulcanization, 160–61

World War II, 1, 8, 10, 14, 40, 51, 109,
  135

X-ray diffraction, 47, 78, 103

Name Index

Abbott Laboratories, 71
Aberdeen, University of, Scotland, 129
Adams, Clark E., 80, 129
Adams, Roger, 70–71
Akron, Ohio, 28, 86. *See also* Brecksville; Government Laboratories
Akron, University of, Ohio, 82, 102, 116–17, 130
Alfrey, Turner, Jr., 128, 131
American Chemical Society: Division of Polymer Chemistry, 124, 134–35; Rubber Division, 46, 87, 134
American Cyanamid, 40
American Society for Testing and Materials (ASTM), 111
American Telephone & Telegraph (AT&T), 101, 102
Amsterdam, University of, Netherlands, 103
Arizona, University of, Tucson, 135

Bacon, Reginald G. R., 131
Bailey, William J., 42, 48, 133
Baker, William O., 103, 104, 108–9, 111; collaboration with Peter Debye, 84, 98 n.86; correspondence with other groups, 106, 109, 134; gel studies, 105–8, 115, 130, 157; styrene content of GR-S, 15, 105
Baldwin, Francis P., 50
Baltimore, Maryland, 105
Bamford, Clement H., 132
Bancroft, Wilder D., 83

Barnes, R. Bowling, 40
Baruch, Bernard M., 11, 55
Baruch Committee, 11–12, 55, 109–10
Baton Rouge, Louisiana, 35, 50, 114
Batt Committee, 3
Baytown, Texas, 19, 50
Beckman, Arnold O., 40
Bekkedahl, Norman, 109
Bell Telephone Laboratories, 42, 67 n.131, 101, 102–9, 115; collaboration with Peter Debye, 84, 98 n.86. *See also* Baker, William O.; Williams, Robert R.
Berlin, Germany, 54
Biggs, Burnhard S., 103, 104
Billmeyer, Fred W., Jr., 85, 86, 124
Bisio, Attilio, 141
Blackley, nr. Manchester, England, 131
Blake, John T., 87
Borders, Alvin M., 16, 30
Boss, A. Evan, 13, 16
Brattain, Robert R., 40
Brazil, 1, 23
Brecksville, Ohio, 28
Briggs, Lyman J., 109
Britain, 23, 104, 116, 129, 131–32
British Rubber Producers' Research Association, 117, 130
Brooklyn Polytechnic, New York City, 124, 128, 133, 135, 136, 141
Brown, Harold P., 37
Brown, Herbert C., 76
Brownell, Herbert, Jr., 19

Buckley, Oliver E., 84, 97–98 n.86
Bueche, Arthur M. (Art), 85
Bueche, Frederick (Fritz), 85
Burke, Oliver W., 16
Buswell, Arthur M., 7
Buswell, Robert J., 71
Butler, George B., 133

California, 102, 106, 117. *See also*
    Torrance
California Institute of Technology, Pasa-
    dena, 102
Canada, 29, 51. *See also* Sarnia
Canadian Synthetic. *See* Polymer
    Corporation
Carlin, Robert B., 77
Carter, James E., 2
Catalina Island, California, 102
Charles University. *See* Prague, Univer-
    sity of
Chattanooga, University of, Chattanooga,
    Tennessee, 124, 133
Chemical Warfare Service. *See* U.S.
    Army
Chemische Werke Hüls, 22
Chicago, Illinois, 86
Chicago, University of, Illinois, 13, 15,
    16, 61 n.32, 70, 75–79, 103
Cincinnati, University of, Ohio, 15, 30,
    35, 81
Clifford, Albert M., 16
Coca-Cola Co., 71
Cohen, Ernst, 95 n.61
Collyer, John L., 41
Columbia University, New York City,
    130
Compton, Karl T., 11
Conant, James B., 11
Condon, Edward U., 117
Connecticut, University of, Storrs, 129
Copolymer Corporation, 22, 35, 114
Cornell University, Ithaca, New York,
    16, 83–86, 103, 132, 133–34, 135, 136
Corrin, Myron L., 78, 135
Council for National Defense, Advisory
    Committee of, 9
Courtaulds, 132
Cox, John T., Jr., 3–4, 117–18
Cuba, 11
Czechoslovakia, 54, 80

Dale, Wesley J., 80, 81
Davis, Carroll C., 87
Deanin, Rudolph, 109
Debye, Peter J. W., 16, 83–85, 89, 97–98,
    n.86, 127
Derick, Clarence G., 71
Deutschman, Archie, 135
Dewey, Col. Bradley, 12, 52, 53, 101, 105
Dewey & Almy (now part of W. R.
    Grace), 12
D'Ianni, James D., 16, 37, 44, 74, 87
Dinsmore, Ray P., 52, 66 n.122, 101, 115
Dow Chemical Co., 5 n.1, 8, 51, 75
Dunbrook, Raymond F., 13, 16, 87
Dunlop Canada, 22
du Pont de Nemours & Co., E. I., 33,
    53, 91, 103; collaboration with Illi-
    nois, 71; Flory at, 76, 130; history of,
    67 n.131; as "outsider," 51; synthetic
    rubbers, 7–8, 22, 33, 42, 51
Dutch East Indies, 1, 54

Eastman Kodak, 129
Einstein, Albert, 84
Elm, Adolf C., 135
Emeryville, California, 40
England. *See* Britain
Erickson, Charles L., 103
Evans, Meredith Gwynne, 129, 131, 138
    n.22
Ewart, Roswell H., 78

Farwell Cafe, Urbana, Illinois, 72
Federal Facilities Corporation, 47
Feldon, Milton, 113
Fenton, Henry J. H., 97 n.80
Ferry, John D., 133
Field, John E., 40, 41, 84
Fikentscher, Hans, 129
Firestone, Harvey, Jr., 29
Firestone Tire and Rubber, 22, 76, 90,
    117; collaboration with Jersey Stan-
    dard, 9, 51; employees co-opted to
    RRC/ORR, 13, 16; history of, 4 n.1;
    sale of Government Laboratories to,
    21–22, 116; synthetic "natural rub-
    ber," 44, 46–47, 49, 58, 64–65 n.100
Fitch, Robert M., 129
Florida, University of, Gainesville, 133,
    136
Florida State University, Tallahassee, 133

Flory, Paul J.: collaboration with W. O.
Baker, 106; at Cornell, 83, 85–86,
132–34; free radical polymerization,
76, 77; gelation, 130; at Goodyear,
128, 130, 131, 132; infra-red spectro-
photometry, 40; polymer configura-
tions, 85–86; publications, 135;
rubber networks, 131
Fly, James Lawrence, 97 n.86
Foster, Fred C., 64–65 n.100
Fox, Thomas G., 86, 132
France, 23
Francis, Clarence, 9
Francis Committee, 9
Frank, Robert L., 31, 72, 104
Friedman, Les A., 108
Fryling, Charles F., 17, 35
Fuller, Calvin S., 12–13, 52, 66 n.122, 98
n.86, 103, 104, 115, 134. *See also*
Williams, Robert R.
Fuoss, Raymond M., 133

Gehman, Samuel D., 28, 41, 84, 109,
113, 117
General Electric, 67 n.131
General Tire and Rubber: collaboration
with I. G., 7–8; oil-extended rubber,
17, 37–38, 39, 55; "outsider," 29, 51,
59
George, Kenyon L., 40
Germany, 1, 36, 56, 73, 79, 104, 109,
117; West, 22–23, 45. Germany *and*
German *are usually synonyms for*
I.G. Farben
Gibbs, Carlin F., 45
Gibson Island, Maryland, 87–88
Gibson Island (Gordon) Conferences, 87–
88, 99 n.115, 134–35
Gillette, Guy, 11
Gillette Committee, 11
Gilliland, Edwin R., 52, 66 n.122, 140
n.62
Gilliland Committee, 140 n.62
Goldfinger, George, 128
Goldsmith, Henry, 113
B. F. Goodrich, 16, 22, 28, 29, 30, 51,
131; American Rubber Team, 18, 20,
42–43, 58; bromobutyl, 50; collabora-
tion with Government Laboratories,
18, 39, 114, 116, 135; early synthetic
rubbers, 8–9, 74; Goodrich Flexome-

ter, 113, 117; infra-red spectrophotometry, 63 n.80; low-hysteresis rubbers,
18, 41–43; lukewarmness toward research program, 28–29, 43, 55, 58,
90; oil-extended rubber, 36–37, 39;
redox polymerization, 33, 39, 61
n.32; research expenditure, 53, 66
n.126; synthetic "natural rubber,"
20, 45–46, 46–47, 47–48, 49, 64 n.89
Goodrich-Gulf, 20, 45–46, 47, 49, 58
Goodyear Tire and Rubber, 22, 28, 29,
30, 51; California Development Program, 33–34, 106, 107, 108; collaboration with universities, 90–91, 131;
early synthetic rubber, 8–9, 74; em-
ployees mentioned, 16, 52, 109;
Flory at, 128, 130, 131, 132; Gehman
torsion technique, 113, 117; gel studies at Torrance, 106, 107, 108;
history of, 4 n.1; infra-red spectrophotometry, 40, 41; light scattering,
84; oil-extended rubber, 17, 37–38,
39; synthetic "natural rubber," 20,
44, 46; vinylpyridine copolymer, 74
Gordon, Neil E., 88
Government Laboratories, Akron, Ohio,
102, 112–14, 116–17, 118; collaboration with Goodrich, 18, 39, 114, 135;
collaboration with Illinois, 52, 74–75,
113, 135; evaluation of new polymers, 47, 74–75, 113; sold to Firestone, 21–22, 116
Government Tire Testing Fleet, 35, 110,
113
Greece, 54
Greer, Paul S., 16
Grisdale, Richard O., 120 n.40
Groves, Gen. Leslie R., 53
Gulf Petroleum, 20, 45
Guth, Eugene, 131, 133

Haber, Fritz, 75
Haller, Elden D., 40
Hampton, Robert R., 41
Harkins, William Draper, 13, 15, 75–76,
89, 90; consultant for U.S. Rubber,
77, 91; emulsion polymerization, 16,
78, 129
Harries, Carl D., 73
Harris, Walter E., 80–81
Hart, Edwin J., 41

Harvard University, Cambridge, Massachusetts, 11
Haward, Robert N., 129, 138 n.22
Helin, Arthur F., 113
Herbert, Vernon, 141
Hercules Powder, 22, 32, 34, 51, 81, 133
Heyrovsky, Jaroslav, 80
Hoover, Herbert, 12
Horne, Samuel E., 20, 45, 49
Houdry Process Corporation, 23
Houston, Texas, 22, 38
Houston, University of, Texas, 133
Howe, Clarence D., 29
Howland, Louis H., 33
Humble Oil, 5 n.1, 40, 50

I.G. Farbenindustrie, 22, 42, 74, 129; Buna 85, 36; Buna N, 7–8, 9, 18; Buna S, 7–8, 36, 73, 89, 91, 104; Buna S3, 14, 36; Buna S10, 18; butyl rubber, 49; collaboration with Jersey Standard, 8–9, 12; continuous polymerization, 114; infra-red spectrophotometry, 63 n.77; Leverkusen rubber laboratory, 102, 109; Mark-Wulff process, 75; masterbatching patent, 38; modifier patents, 14; negotiations with Goodrich and Goodyear, 8; oil-extended rubber, 36; redox polymerization, 34–35, 38; thermal degradation of Buna S, 36, 107
Illinois, University of, Urbana, 13, 15, 109; collaboration with Government Laboratories, 21, 52, 74–75, 113, 135; education of Ph.D.'s, 71–72, 133–36, 141; history of, 70–71; inhibitors, 72; modifiers, 31, 72, 104; polymerization kinetics, 77, 80, 128, 129; replacement of styrene by new monomers, 18, 21, 73–75, 113; soap problem, 14, 31; sodium polymerization, 20–21, 42, 47; structure of GR-S, 72–73
Imperial Chemical Industries, 66 n.127, 104, 131
Indochina, French, 10
Indonesia. See Dutch East Indies
Ingmanson, J. H., 104
Institute, West Virginia, 19, 114

Interagency Policy Committee on Rubber. See Batt Committee
Iowa, 11
Italy, 23
Ithaca, New York, 83. See also Cornell University

James, Hubert M., 131
Japan, 1, 2, 9–10, 23
Jeffers, William M., 12, 53
Jersey Standard. See Standard Oil Co. of New Jersey
Jones, Jesse, 10, 102, 109
Juve, Arthur E., 28
Juve, Robert D., 115

Katz, Johan R., 103
Kemp, Archie R., 101, 102–3, 104, 106, 115
Kent State University, Kent, Ohio, 28
Kettering, Charles F., 67 n.131
Kharasch, Morris S., 13, 75; collaboration with U.S. Rubber, 77, 91; correspondence with W. O. Baker, 109, H. formula, 61 n.32; inhibition of popcorn polymer, 15; mechanism of GR-S polymerization, 77, 78; peroxides, 76–77, 78–79; polymerization with lithium hydride, 44, 88; pre-war research, 76–77; publications, 135
Kirkwood, John G., 83
Kolthoff, Izaak M. (Piet): analysis, 15, 79–80, 131; collaboration with H. A. Laitinen, 15, 72; correspondence with W. O. Baker, 109, 134; diazothioethers, 38, 81; key figure in research program, 13, 17, 91; modifiers, 15, 30, 80–81; oxidative analysis of GR-S, 73; pre-war research, 79–80; publications, 135; redox systems, 34–35, 81–83; relationship with rubber companies, 90–91
Konrad, Erich, 109
Korea, South, 17. See also Korean War in subject index
Kranzberg, Melvin, 89
Krejci, Joe C., 32
Krigbaum, William R., 86, 133
Kruyt, Hugo R., 79
Kuhn, Werner, 131

Labbe, B. G., 113
Laitinen, Herbert A., 15, 72
Langlois, Richard N., 116
Langmuir, Irving, 76
Laundrie, Robert W., 113, 114
Leeds, University of, England, 129, 138 n.22
Leeper, Harold, 113
Lehigh University, Bethlehem, Pennsylvania, 133, 136
Leverkusen, nr. Cologne, Germany, 7, 102, 109
Lewis, Frederick M., 128
Lewis, Warren K., 66 n.122
Lind, Samuel C., 79–80
Lingane, James J., 80
Linnig, Frederic J., 112
Livingston, John W., 34, 117–18
Los Angeles, California, 19. *See also* Torrance
Ludwigshafen am Rhein, Germany, 63 n.77, 75. *See also* Oppau
Luft, Karl, 63 n.77

Madorsky, Irving, 111
Madorsky, Samuel L., 112
Maher, E. D., 108
Maidenhead, Berkshire, England, 132
Malaya, 1, 54
Malm, Frank S., 101, 104
Mandel, John, 110
Mandelkern, Leo, 86, 133
Manhattan Project, 53, 59, 67 n.131, 89, 134
Mark, Hans, 117
Mark, Herman F., 71, 75, 131, 133, 135, 141
Marker, Russell E., 76
Marvel, Carl S. (Speed): copolymerization, 128; intelligence activities in Germany, 34; key figure in program, 13, 70, 91; letter from Williams, 104; polybenzimidazole (PBI), 18, 75, 89, 141; polymer education, 133; pre-war research, 71–72; publications, 135; soap research, 31; sodium rubbers, 20–21, 42, 43, 44, 47, 88, 91; structure of rubbers, 73; synthesis of new modifiers, 72, 104; synthesis of new monomers and rubbers, 18, 21, 74–

75, 113; U.S. Air Force–sponsored research, 21, 54
Maryland, University of, College Park, 48
Massachusetts, University of, Amherst, 133
Massachusetts Institute of Technology, Cambridge, 11, 20, 43, 52, 53, 66 n.122, 75, 79, 133–34
Matthews, F. E., 42, 71, 146
Mayer, Joseph, 130
Mayo, Frank R., 28, 76, 77–78, 128–29
McMillan, Frank M., 64 n.100, 65 n.104
McPherson, Archibald T., 109, 110
Medalia, Avrom I., 82–83
Meehan, Edward J., 34, 35, 74, 81, 82, 90
Mellon Institute, Pittsburgh, Pennsylvania, 10
Melville, Sir Harry, 128–29, 131
Messer, William E., 33
Meyer, Albert W., 41
Michigan, University of, Ann Arbor, 40
Midgley, Thomas, Jr., 43
Minnesota, University of, Minneapolis, 13, 15, 30–31, 73, 74, 79–83, 91, 100 n.124
Mooney, Melvin, 28, 106, 111, 130
Morawetz, Herbert, 127
Morgan, Thomas Hunt, 86
Morrissey, Robert T., 50
Morton, Avery A., 20, 43, 44, 88, 90–91
Morton, Maurice, 78, 124, 130
Mullen, James W., II, 106, 107

National Academy of Sciences, 55
National Aeronautics and Space Administration, 59, 137 n.3; Ames Research Center, 117. *See also* Space program *in subject index*
National Bureau of Standards, 41, 101–2, 109–12, 113, 115, 117; measurement of styrene content of GR-S, 15, 104, 105, 111
National Institutes of Health, 136
National Science Foundation, 21–22, 70, 134, 137, 140 n.62; Mohole project, 127
National Technical Laboratories, Inc., 40, 81, 127
Natta, Giulio, 88
Naugatuck, Connecticut, 61 n.32

Nelson, Donald M., 10, 12, 110
Nelson, Richard R., 116
Netherlands, 79
New Hampshire, 87
New York City, 86
New York University, New York City, 112
Newhall, Arthur B., 10, 110
Nixon, Richard M., 59
Northwestern University, Evanston, Illinois, 133
Notre Dame, University of, Notre Dame, Indiana, 18, 51, 131, 133
Noyes, Arthur A., 75
Noyes, William A., Sr., 71
Nudenberg, Walter, 77, 78–79

Office of Naval Research. *See* U.S. Navy
Office of Production Management, 10
Office of Rubber Reserve. *See* Rubber Reserve, Office of
Office of Scientific Research and Development (OSRD), 53
Office of Synthetic Rubber. *See under* Reconstruction Finance Corporation
O'Neil, William, 37
Oppau, nr. Ludwigshafen, Germany, 49
Osterhof, H. Judson, 30, 37
Ostromislensky, Ivan I., 23 n.2

Passaic, New Jersey, 77
Pearl Harbor, Hawaii, 10
Pearson, Gerald L., 66 n.122
Pennsylvania, University of, Philadelphia, 133
Perkin-Elmer, 73
Perry, Charles W., 107
Peters, Henry, 103, 104, 106
Petrochemicals Ltd, 129
Petroleum Administration for War, 40
Pfau, Ernest S., 37, 39
Philadelphia, Pennsylvania, 46, 103
Phillips Petroleum, 40; carbon black, 32; cold rubber, 17, 35, 82; collaboration with Goodyear, 8, 51; history of, 5 n.1; modifiers, 15, 30, 31; "outsider," 29, 51, 58; synthetic "natural rubber," 20, 46
Poland, 109
Polymer Corporation, 29, 37, 39, 50, 51, 82, 108

Polymer Research Discussion Group, 51, 86–88, 124, 132, 134–36
Porter, Paul A., 98 n.86
Prague, University of, 80
President's Science Advisory Committee. *See* Gilliland Committee
Price, Charles C., 18, 51, 77, 80, 89, 128, 129, 133
Priest, William J., 129
Princeton University, Princeton, New Jersey, 103, 133
Proctor and Gamble, 14, 31
Pummerer, Rudolf, 73

Rabjohn, Norman, 73, 74
Rabkin, Yakov M., 127
Rayleigh, 3rd Baron (John W. Strutt), 84
Reconstruction Finance Corporation (RFC), 2, 53, 135; cold rubber, 17, 35, 39, 57; disputes with Goodrich and Firestone about synthetic "natural rubber," 45–46, 46–47, 49; Office of Synthetic Rubber, 12, 47; oil-extended rubber, 37; plant disposal plan, 19; research contracts, 29; RRC subsumed into, 12; shift of research from RFC to military research offices, 18, 21. *See also* Rubber Reserve, Office of
Rehner, John, 131
Reich, Murray H., 114
Reilly Tar and Chemical Co., 74
Research Corporation, 66 n.122
Resinous Products and Chemical Co., 103
Reynolds, William B., 15, 17, 30, 31, 35, 38, 81, 90
Rickover, Hyman, 53
Roe, Charles P., 85
Rohm and Haas, 103, 113
Romney, George W., 59
Roosevelt, Franklin D., 9, 10, 11
Rossini, Frederick D., 111
Rothe, F. L., 112
Rounds, Leslie R., 19
Rounds Commission, 19
Rubber Director, Office of the (ORD), 12, 13, 31, 52–53, 70, 105, 106, 108; Committee on Revision of GR-S Formula, 32–33; Copolymer Equipment Development Branch, 107; Polymer Research Policy Committee, 87,

116; Research Branch, 12–13, 118
n.4; Rubber Research Board, 34, 70,
131; Subcommittee on Polymer
Structure, 132; Temporary Committee on Standardized Research Meth-
ods, 132. *See also* Polymer Research
Discussion Group
Rubber Division. *See under* American
Chemical Society
Rubber Producing Facilities Disposal
Commission. *See* Rounds
Commission
Rubber Reserve, Office of (ORR), 12,
111, 112, 132; Government Labora-
tories and, 112, 114, 117; Research
and Development Branch, 13, 16, 90,
109; Subcommittee on Test Methods,
111, 113
Rubber Reserve Company (RRC), 9–13,
29, 31, 33, 50, 72, 102, 108, 124, 130;
Operators' Committee, 34; Polymer
Research Branch, 108; Technical
Committee, 87
Rubber Survey Committee. *See* Baruch
Committee.
Ruhrchemie, 45
Russia. *See* Soviet Union

Sarnia, Ontario, Canada, 37, 50, 108
Scheraga, Harold A., 86
School, Nicholaas, 79
Schroeder, Herman E., 42
Scott Testers Inc., 112
Sebrell, Lorin B., 8
Semon, Waldo L., 8, 18, 45, 46
Semperit, 36
Seymour, Raymond B., 133
Shell Chemical Corporation, 20, 46
Shell Development Co., 5 n.1; 8, 40, 51,
127
Shepard, Morris G., 16
Shipman, James J., 45
Shire, J., 120 n.40
Shockley, William, 108
Siemens, Werner, 102
Simha, Robert, 112
Smith, Paul V., 31
Smith, Wendell V., 78
Smyth, Charles P., 103
Solo, Robert, 3–4, 39, 52, 54–55, 59
Soviet Union, 16, 17, 18, 54, 104

Sparks, William J., 49
Special Committee of the U.S. Senate In-
vestigating the National Defense Pro-
gram. *See* Truman Committee
Standard Oil Company of New Jersey, 8–
9, 10–11, 39, 40, 51; butyl rubber,
49–50; Flory at, 40, 131, 132; history
of, 5 n.1
Starkweather, Howard, Sr., 33
Staudinger, Hermann, 71, 84, 106
Stavely, Frederick W., 20, 44, 46, 49, 58,
76
Stein, Richard S., 133
Stewart, William D., 33
Stieglitz, Julius, 76
Stiehler, Robert D., 110
Stockmayer, Walter H., 130, 133–34
Street, John N., 16
Sun Oil, 67 n.131
Swan, Thomas, 16

Taft, William K., 113, 114
Taylor, Rolla H., 111
Texaco, 22
Texas, University of, Austin, 103
Texas-US Chemicals, 22
Thomas, Robert M., 49, 50
Tobolsky, Arthur V., 133
Torrance, nr. Los Angeles, California,
33–34, 106, 107, 108
Truman, Harry S, 3
Truman Committee, 41, 63 n.81
Trumbull, Harlan L., 16, 90
Tsai, C. H., 129

Union Carbide, 11
Union Pacific Railroad, 12
United Carbon, 19
United States Air Force, 21, 54, 136
United States Army: Chemical Warfare
Service, 112; Gas-Flame Division, 76
United States Congress, 11, 19, 41, 51, 53
United States Department of: Agriculture,
21, 113, 114, 135; Commerce, 10, 102,
109; Defense, 18; Justice, 10, 19
United States Navy: Allegheny Ballistics
Laboratory, 133; Naval Stores Re-
search Division, 114; Office of Naval
Research, 48, 66 n.120, 131, 132, 136

U.S. Rubber (Uniroyal), 9, 16, 22, 28, 38, 51, 61 n.32, 82; fundamental research, 77–78, 78, 85, 91, 128–29; gel research, 106, 107, 130; history of, 4 n.1; infra-red spectrophotometry, 41, 47; low temperature polymerization, 33, 34; Mayo's research group, 28, 77–78, 128–29; Mooney viscometer, 28, 117, 130

U.S. Senate Committee on Agriculture and Forestry, subcommittee on synthetic rubber. *See* Gillette Committee

Universal Oil Products, 75

Utrecht, University of, Netherlands, 79, 95 n.61

Vienna, University of, Austria, 131

Vila, George R., 106

Visscher, Maurice, 90, 100 n.124

Wakefield, Lynn B. 44

Wall, Frederick T., 128, 131, 135

Wall, Leo A., 112

Walling, Cheves, 44, 76, 77, 128, 141

War Production Board (WPB), 10, 11, 12, 110

Washington, D.C., 86, 104, 109

Weidlein, Edward R., Jr., 10

Western Electric, 12, 102

Westheimer, Frank H., 76

Whitby, George Stafford, 82, 87, 102, 114

White, James L., 130–31

White, John U., 40

White, Leland M., 107, 115

Whitney, Willis R., 67 n.131

Williams, Robert R., 51, 66 n.120; aims for Government Laboratories, 118 n.4; good research leader, 12–13, 52–53; research leader at BTL, 102, 104–5; setting up the rubber research program, 12–13, 79, 115; "Summary of Present Research," by Williams and C. S. Fuller, 52, 72, 73, 86; support for activated recipe, 32

Wisconsin, University of, Madison, 133

Wood, Lawrence A., 109–10, 111, 112

Woodford, D. E., 41

Wulff, Carl, 75

Yale University, New Haven, Connecticut, 133, 136

Zelinski, Robert P., 20

Ziegler, Karl, 20, 43, 45–46, 49, 88 , 89, 99 n.118

Zwicker, Benjamin M. G., 30

Polymers and Monomers Index

Acrylate, ethyl: copolymer with chloroethyl vinyl ether, 113
Acrylic acid: monomer, 113; copolymers
with butadiene, 42, 74
Acrylonitrile: polymer, 132; copolymer
with butadiene, 7–8, 9, 21, 33, 40,
113
Adiprene (Du Pont polyurethane rubber), 42
Ameripol (Goodrich butadiene-
methacrylate copolymer), 9
Ameripol SN (Goodrich synthetic "natural rubber"), 20

Benzalacetophenone: copolymer with butadiene, 75
Bromobutyl, 50, 51
Budene (Goodyear all-*cis* polybutadiene),
20
Buna 85 (I.G. polybutadiene), 36
Buna N (I.G. butadiene-acrylonitrile copolymer), 7–8, 9, 18
Buna S (I.G. butadiene-styrene copolymer), 7–8, 36, 73, 89, 91, 104; Buna
S3, 14, 36; Buna S10, 18
Butadiene: as a component of GR-S, 1,
7, 21, 73, 127, 153; monomer, 8–9,
22, 49, 146; polymer, 20–21, 36, 41,
42–44, 47, 73, 78, 146–48, 153–59;
stereoregular polymer, 20, 23, 46, 51;
production of butadiene from petroleum, 11, 14; production of butadiene from alcohol, 11, 13–14, 17,

19. *See also* Popcorn polymer; *and
under the other comonomer for butadiene copolymers*
Butyl rubber, 39, 49–50, 159
Butylene: monomer, 146
Butylene, iso-: monomer, 159–60; polymer, 40, 49, 159

Chemigum (Goodyear butadiene-
acrylonitrile copolymer), 9
Chlorobutyl, 50
Chlorostyrene: monomer, 74; *ortho-*
chlorostyrene copolymer with butadiene, 73; *para*-chlorostyrene polymer, 85
Cinnamate, methyl: polymer, 75
Cinnamic acid: copolymer with butadiene, 75

Duprene (Du Pont polychlorobutadiene,
renamed Neoprene in 1937), 7

Estane (Goodrich polyurethane rubber),
42
Ethylene: monomer, 144–45; polymer,
20, 45, 88, 104, 112, 145, 151; copolymer with isoprene, 20, 45; copolymer
with isopropenyl acetate, 48; co/ter-
polymers with propylene, 22, 51

Fluorocarbon rubbers, 18, 51

GR-S, *passim*; GR-S latex, 82; high-styrene resins, 51; "rapid" GR-S, 18, 114; GR-S 60, 130. *See also* Rubber, cold
Gutta-percha, 102

Hycar PA 21 (Goodrich ethyl acrylate and chloroethyl vinyl ether copolymer), 113
Hycar 2202 (Goodrich bromobutyl), 50

Isoprene: monomer, 46, 49, 102, 109, 146–47; polymer, 43, 58, 147–48; stereoregular polymer: synthetic "natural rubber," 20, 23, 42, 44–46, 51, 92; synthetic "balata," 51. *See also* Rubber, natural

Lactoprene EV (USDA ethyl acrylate and chloroethyl vinyl ether copolymer), 113

Methacrylate, methyl: polymer, 85, 112, 129; copolymers with butadiene, 9, 42; copolymer with styrene, 129; terpolymer with isoprene and butadiene, 42

Natsyn (Goodyear synthetic "natural rubber"), 20
Neoprene (Du Pont polychlorobutadiene), 7
Nirub C (experimental Goodrich terpolymer with isoprene and butadiene), 42

Paracon (BTL polyester from ethylene glycol, and sebacic and succinic acids), 103
Paraplex (Resinous Products' name for Paracon), 103
Polyamides, 103, 105, 132
Polybenzimidazole (PBI), 18, 75, 89, 141
Polyelectrolytes, 132
Polyesters, 103, 130, 132, 148–49; preparation of polyesters from citraconic acid and mucaconic acid by Goodrich, 48
Polysulfones, 71
Polyurethanes, 42, 89

Popcorn polymer (polybutadiene): formation of, 15, 78
Propylene: stereoregular polymer, 88–89; co/terpolymers with ethylene, 22, 51
Propylene oxide: polymer, 51; polyurethane from, 89

Rubber, cold (GR-S prepared at low temperatures): development of: by Akron (University of), 82; by Chicago, 77, 79; by Goodrich and Government Laboratories, 21, 39, 114; by Minnesota, 34–35, 81–83; by Phillips Petroleum, 17, 35, 51, 82; by Polymer Corp., 29, 51, 82; an achievement of the research program, 41, 55, 141; development discussed, 38–39, 57–58, 89, 91–92; early research, 32–34; introduced by RFC, 17, 35; production of, 17, 19, 35
Rubber, methyl (polymethylisoprene), 79
Rubber, natural, 1, 8, 17, 38, 42, 84, 102–3, 109, 111, 148; reclaim, 110. *See under* Isoprene *for synthetic "natural rubber"*
Rubber, oil-extended, 17–18, 19, 29, 36–38, 39, 51, 56
Rubbers, heat-resistant, 18, 21, 75
Rubbers, low temperature-resistant ("arctic rubbers"), 18, 21, 54, 89, 113
Rubbers, oil-resistant, 7–8, 18, 54, 74, 113

Shell Isoprene Rubber (Shell synthetic "natural rubber"), 20
Silicones, 14, 18, 113, 132
Styrene: as a component of GR-S, 1, 7, 18, 21, 27, 73–75, 127, 153; monomer, 14, 19, 152–53; polymer, 77–78, 80, 112, 129; copolymer with methyl methacrylate, 129; sodium-polymerized copolymer with butadiene, 42. *See also* Styrene content, measurement of *in subject index*

Thiokol (polysulfide rubber), 110

Vinyl chloride: polymer, 71, 85; copolymer with vinyl acetate, 128
Vinyl pyridine: polymer, 85; copolymer with butadiene, 74